THE

SELF-ORGANIZING ECONOMY

Paul Krugman

BLACKWELL
Publishers

First published 1996

Blackwell Publishers, Inc.
238 Main Street
Cambridge, Massachusetts 02142
USA

Blackwell Publishers, Ltd.
108 Cowley Road
Oxford OX4 1JF
UK

Library of Congress Cataloging–in–Publication Data
   The self-organizing economy / Paul R. Krugman.
      p.   cm.
   Includes bibliographical references and index.
   ISBN 1-55786-698-8 (hc)
   ISBN 1-55786-699-6 (pb)
      1. Economics.   2. Self-organizing systems.   I. Title.
   HB199.k75   1995
   338.9--dc20   95-31593
                                                                      CIP

British Library Cataloguing in Publication Data

A CIP catalogue record for this book is available from the British Library.

Composition by Megan H. Zuckerman

Printed in the USA by Book Crafters

This book is printed on acid-free paper

Contents

Preface

Every once in a while the world economy plunges into a severe recession. Some of these international slumps appear to be caused by particular noneconomic events: wars, disruptions of oil supply. Others, however, have no obvious cause – and their global scale is hard to explain in terms of the conventionally measured linkages among national economies.

Every once in a (much longer) while, the paleontologists tell us, the world experiences mass extinctions that wipe out most of the extant species. Some extinctions appear to have external causes, like the comet whose impact coincided with the demise of the dinosaurs. Others, however, have not been tied to any obvious cause. Some theorists who simulate evolution on their computers claim that this is as it should be: their models predict occasional, spontaneous mass extinctions even in the absence of any external shocks to the system.

Can these two paragraphs have anything to do with each other? Is there a sense in which a global slump is something like a mass extinction?

Here is another parallel. When you look at the *sizes* (however measured) of many enormously complex physical or biological phenomena, the distribution of those sizes for some reason turns out to be well described by a very simple *power law*: the number of objects (earthquakes, meteorites, species, and – perhaps – extinctions) whose size exceeds S is proportional to S^{-a}, where a is not only a mystery parameter but often turns out, weirdly, to be a round number, like 1 or 2. Among the most spectacular examples of a power law, however, is one that involves economics rather than physical science: the size distribution of cities. In the United States, the number of cities whose population exceeds S is, simply, proportional to $1/S$: there are 40 cities with more than a million people, 20 with more than 2 million, and 9 (Houston fell a bit short) with more than 4 million!

Social scientists are normally suspicious of people who want to import concepts from physical or biological science, and with good reason: the history of such efforts, from social Darwinism to systems dynamics, has been little short of disastrous. Nonetheless, this time things may be different; there is a genuinely interesting interdisciplinary movement of which economics ought to be a part.

In the last few years the concept of *self-organizing systems* – of complex systems in which randomness and chaos seem spontaneously to evolve into unexpected order – has become an increasingly influential idea that links together researchers in many fields, from artificial intelligence to chemistry, from evolution to geology. For whatever reason, however, this movement has so far largely passed economic theory by. It is time to see how the new ideas can use fully be applied to that immensely complex, but indisputably self-organizing system we call *the economy*.

In this book I try to show how models of self-organization can be applied to many economic phenomena – how the principle of "order from instability," which explains the growth of hurricanes and embryos, can also explain the formation of cities and business cycles; how the principle of "order from random growth" can explain the strangely simple rules that describe the sizes of earthquakes, meteorites, and metropolitan areas. I believe that the ideas of self-organization theory can add substantially to our understanding of the economy; whatever their ultimate usefulness, these ideas are very exciting, and playing around with them is tremendous fun.

Finally, a note on style. This book began as the Mitsui lectures, which I gave at the University of Birmingham in March 1994. Although I have not tried to maintain the lecture format, I have allowed myself to retain some of the license usually granted in such a lecture series: this book is written in an informal style and contains more than a few wild speculations. Nonetheless, I have tried to get things right when I can; and I am grateful for discussions with participants in seminars at Birmingham, UCLA, Chicago, and Stanford that helped me correct some serious errors. In particular, I would like to thank Mike Woodford at Chicago for pushing me to test a pet hypothesis to its well-deserved destruction.

PART
1

Embryos, Earthquakes, and Economics

Is an economic slump like a hurricane, or is it more like an earthquake? Is a growing city like an embryo, or is it more like a meteorite?

These questions may sound like children's riddles, to which the answer is some kind of bad pun; or maybe they sound like Zen koans along the lines of "What is the sound of one hand clapping?" in which the absurdity of the question is meant to jolt the listener into a higher state of awareness. But my aim is neither to tell jokes nor to help you achieve an enlightenment that transcends rationality. On the contrary, in the course of this book I hope to convince you that these are perfectly reasonable questions to ask, that there may well be a more than poetic sense in which a deepening slump resembles an emerging hurricane, in which a growing city is quite a lot like a developing embryo.

In making these kinds of analogies I am not, of course, being completely original. There is a broad and growing interdisciplinary movement in the physical and biological sciences – often referred to as the study of *complexity* – that looks for exactly such parallels between seemingly disparate phenomena. For example, in the course of my preparations for this book I ran across an article on "percolation theory" that casually listed 15 areas to which the basic approach applied, ranging from atomic nuclei to galaxies. But so far this movement has largely passed economics by.

Actually, let me qualify that statement. People who write books or convene conferences on complexity almost invariably assert that the emerging field will make great contributions to the study of

1

that immensely complicated system we call the economy. Indeed, the Santa Fe Institute, one of the hotbeds of this style of research, was initially funded by Citibank largely because John Reed, the bank's CEO, hoped that research on complex systems would improve economic forecasting. For whatever reason, however, the authors of articles and books on complexity almost never talk to serious economists or read what serious economists write; as a result, claims about the applicability of the new ideas to economics are usually coupled with statements about how economies work (and what economists know) that seem so ill-informed as to make any economist who happens to encounter them dismiss the whole enterprise.

But it does not have to be that way. What I am going to claim, and I hope demonstrate, in these lectures is that some of the ideas that come out of the interdisciplinary study of complex systems – the attempt to find common principles that apply across a wide variety of scientific fields, from neuroscience to condensed matter physics – are, in fact, useful in economics as well. That is, you can understand and respect the economic theory we already have and still find ways both to improve it and to build bridges to other fields by taking into account the ideas of these interdisciplinary theorists.

Before I go any further, I had better explain a little better what I am talking about. What is this interdisciplinary effort that I have been alluding to, and how can it teach economists anything new?

Many of the people who think that earthquakes, embryos, cycles, and cities have all got something to do with one another describe their field as the study of "complexity," based on the insight that complicated feedback systems have surprising properties. If that were all there were to the program, however, economists would be entitled to change the channel. If economists do understand one thing much better than the lay public, it is the sheer complexity of the economic system and the importance of feedbacks. After all, what is general equilibrium theory but a formalization of the proposition that everything in the economy affects everything else, in at least two ways? If you have ever tried to explain to a roomful of engineers why higher manufacturing productivity will probably reduce, not increase, manufacturing employment – and why higher productivity will not necessarily reduce the trade deficit – you quickly realize that economists are better, not worse, than most physical scientists at understanding the importance of feedback in complex systems.

A second definition, by Philip Anderson, the Nobel laureate physicist who may perhaps be regarded as the father of the field, is that complexity is the science of "emergence." That is, it is about how large interacting ensembles – where the units may be water molecules, neurons, magnetic dipoles, or consumers – exhibit collective behavior that is very different from anything you might have expected from simply scaling up the behavior of the individual units. (The behavior may also be oddly similar to that of ensembles of otherwise very different units: collections of neurons may behave a lot like collections of magnetic dipoles.) Anderson's prime example of an emergent property is the liquidness of water, which is in no sense an extrapolation of some primordial liquidness of individual water molecules.

Here again, however, we have a definition that sounds like what economists already understand pretty well. When Adam Smith wrote of the way that markets lead their participants, "as if by an invisible hand," to outcomes that nobody intended, what was he describing but an emergent property? And examples of emergence abound in economic theory – we need only note the way that competitive markets, in which each individual is striving only for his or her own profit, act as if the participants were collectively trying to maximize the sum of consumer and producer surplus, concepts of which they are generally unaware. (Nor is this only theory: experimental markets in which the participants are assigned payoffs and then make bids for and offers of units of a notional commodity do, in practice, come very close to maximizing aggregate surplus, even though the participants not only are not trying to achieve that objective, they do not even know what each others' payoffs are.)

There is, however, a third definition of this field that does not sound like something that economists already do, or at least not what they do routinely. This is the view that what links the study of embryos and hurricanes, of magnetic materials and collections of neurons, is that they are all *self-organizing systems*: systems that, even when they start from an almost homogeneous or almost random state, spontaneously form large-scale patterns. One day the air over a particular patch of tropical ocean is no different in behavior from the air over any other patch; maybe the pressure is a bit lower, but the difference is nothing dramatic. Over the course of the next few days, however, that slight dip in pressure becomes magnified through a process of self-reinforcement: rising air pulls water vapor up to an altitude at which it condenses, releasing heat that reduces the pressure further and makes more air rise, until that particular piece of the atmosphere has become a huge, spinning vortex. Early in the process of growth an embryo is a collection of nearly identical cells, but (or at least so many biologists believe) these cells communicate with each other through subtle chemical signals that reinforce and inhibit each other, leading to the "decision" of some cells to become parts of a wing, others parts of a leg.

Is the economy a self-organizing system in this sense? Of course it is. Think about a metropolitan area – even, or rather especially, a modern metropolitan area like greater Los Angeles, with no clearly defined center. Is an urban sprawl like LA a homogeneous, undifferentiated mass? No – it is a patchwork of areas of very distinct character, ranging from Koreatown to Hollywood, Watts to Beverly Hills. And it contains (according to the recent book *Edge City* by Joel Garreau) no fewer than 16 "edge cities," newly emerged business centers, each of which includes at least 5 million square feet of office space and 20,000 workers, where the low-rise sprawl suddenly gives way to tall buildings and multistory parking garages. What is so striking about this differentiation is that it is so independent of physical geography: there are no rivers to set boundaries, no big downtown to define a gradient of accessibility. (OK, the beaches and the freeways create a bit of exogenous structure, but grant me a little license.) The strong organization of space within metro Los Angeles is clearly something that has emerged, not because of any inherent qualities of different sites, but rather through self-reinforcing processes: Koreans move to Koreatown to be with Koreans, beautiful people move to Beverly Hills to be with other beautiful people. And when an observer like Garreau traces the development of edge cities, he is immediately drawn to metaphors like "spontaneous combustion" and "critical mass" – clear indicators that he is trying to describe a self-organizing system.

I have started with the way that cities spontaneously evolve a pattern of sharply distinct districts, because the economic self-organization of space is something we can all relate to our immediate experience. But I would assert that there are processes in the economy that produce *temporal* self-organization as well. I refer to the business cycle: the pulses of expansion and contraction around a relatively stable long-run trend. (No, I don't believe in Kondratieff waves.) Some recessions and recoveries are clearly set off by specific, essentially exogenous events like oil crises. Over the long sweep of history, however, most booms and slumps have had no obvious external cause. Most notably, the mother of all economic slumps, the contraction from 1929 to 1933, came as it were out of a clear blue sky. But then, so do hurricanes.

In fact, let me explain why a slump may be quite a lot like a hurricane. A hurricane is a self-reinforcing process, in which an updraft that pulls water vapor to a level at which it condenses thereby releases heat, which in turn reinforces the updraft. A slump is also self-reinforcing: falling output causes firms to slash their investment and consumers to reduce their spending, thereby reducing output still further. But wait, there is more. A hurricane cannot go on forever. The fuel for its violence is the water vapor that exists in relative abundance above a warm tropical ocean. But the hurricane itself cools that ocean surface – indeed, in a way the whole point of tropical storms is that they are Nature's way of transferring solar energy from the tropics to higher latitudes. And so hurricanes, even though they are self-reinforcing in the short run, are self-limiting in the long run. What about an economic slump? Well, it leads to falling prices, or at least disinflation, which gradually increases the real money supply. The low or negative net investment in a deep slump may also lead to a growing backlog of potentially profitable projects. And so we may argue that an economic slump is also a self-limiting process, even if there is no deliberate or effective government policy to get the economy moving again.

By the way, Americans who watch the weather news know that hurricanes sustain themselves by moving, gaining strength when they pass over fresh tracts of warm water; if this has any analogy in business cycles, I am unaware of it. But I did not claim that the parallel was perfect. And you may argue that my description of what goes on in a typical recession is not too accurate – or maybe there is no such thing as a "typical" recession, in which case recessions are more like earthquakes than like hurricanes. But we shall get to that in Part II.

While I am making asides, let me also take the opportunity to make another point: self-organization is not necessarily, or even presumptively, a good thing. I think it is fair to accuse many of the writers on complexity, especially but not only the more popular ones, of falling into this fallacy. Book titles like *Order out of Chaos* (by Nobel laureate Ilya Prigogine and I. Stenger, but with a Foreword by, believe it or not, Alvin Toffler), or *Complexity: Life at the Edge of Chaos* (by R. Lewin) come perilously close to making self-organization a kind of mystical goal. Even Stuart Kauffman, whose *The Origins of Order* is a serious and stimulating tract (if we ignore the inevitably careless and ill-informed section on economics), talks far too casually about coevolutionary systems that maximize "average fitness" – a surely meaningless concept when the fitness of each species is defined at least partly in terms of how well it copes with competition and predation from others. Luckily, if we are at all serious about the economics of self-organization we immediately realize that no value judgment is implied. An economy with a strong business cycle exhibits more temporal self-organization than an economy that grows smoothly, but most of us would rather live in the latter. A city whose racially integrated communities unravel, producing huge segregated domains, becomes more spatially organized, but not better, in the process. Self-organization is something we observe and try to understand, not necessarily something we want.

I have asserted that the study of self-organization is something that economists do not do, or at least do not do routinely. What I have just said about economic slumps is, however, not at all new or original. On the contrary, I have just given you a loose version of the nonlinear business cycle literature that flourished in the 1940s and 1950s, with contributors including economists of the stature of John Hicks, Richard Goodwin, and James Tobin. For that matter, urban economists have hardly been unaware of the self-organization of metropolitan areas, and there have been some notable efforts to model the creation of urban subcenters, as edge cities are colorlessly known among the professionals. (I have in mind particularly the pioneering work of Fujita and Ogawa.)

And yet I think it is fair to say that few economists have explicitly realized that they were trying to model self-organization; few have tried to draw the parallel between self-organization in space and self-organization in time; few have tried to use some of the techniques for understanding self-organization that have evolved in other fields. Furthermore, although some deeply insightful thinkers more or less consciously have written about self-organization in the economy, their work has been neglected by the profession (and is completely unknown outside of it). Nonlinear business cycle theory, for example, was technically far ahead of its time and makes the later efforts to shoehorn "catastrophe theory" into economics look primitive. But who remembers it? (I may be the only economist in my generation who has even heard of it.) Thomas Schelling wrote a remarkable essay on the dynamics of segregation in his wonderful 1978 book *Micromotives and Macrobehavior*, but the book had little impact at the time and is still underappreciated.

I think that I can explain why the pioneers of self-organization in economics have been either neglected or forgotten, and I will talk about those reasons later in these lectures. For now, however, let us put intellectual history aside and turn to doing some economics.

Here is my plan of action. In this first part, I am going to do a quick run-through of stories — I do not want to dignify them by calling them models — about ways in which the economy organizes itself in space. (I start with spatial self-organization for two main reasons. First, as an empirical matter I am on more solid ground: the self-organizing spatial character of the economy is obvious to everyone, although not usually under that name. Second, I know what I believe about spatial economics, but I am still fairly agnostic about the macroeconomics of business cycles.) Along the way I shall describe two broad, and seemingly paradoxical, principles of self-organization that are suggested by these stories and that are also common in much of the literature on self-organization in other fields. Just to whet your appetite, let me give these principles names: "order from instability" and "order from random growth."

In the second part, I will start by showing how these principles may apply not only to space but to time: in particular, how we might think of the business cycle as temporal self-organization. Then we shall come back to the spatial economy. I will describe two more or less full-fledged models that illustrate the principle of order from instability and one that illustrates the principle of order from random growth. Even there I will go light on the equations, stressing an intuitive explanation. But there will be a technical appendix, which derives the results in all the gory detail I can manage (which is not much).

So let us begin our tour with a look at the way that economies organize space, particularly the space within metropolitan areas.

1

Self-Organization in Space

THE VON THÜNEN–MILLS MODEL

How do economists routinely deal with the question of how the economy organizes its use of space? The short answer is that mostly they do not deal with the question at all. Indeed, there is something strange about the way that most of our profession neglects anything having to do with where economic activities happen. For example, the very popular (and almost 900-page-long) economics principles textbook by William Baumol and Alan Blinder contains not a single reference to "cities," "location," or "space" in its index. The rival, 1100-page text by Joseph Stiglitz does contain one reference to cities, which turns out to occur in a brief discussion of rural–urban migration in less-developed countries. Considering how much time most people spend in traffic jams, how many fortunes are made and lost in real estate, the way that we turn a blind eye to spatial economics is little short of eerie. I strongly suspect that this neglect is closely related to the theme of this book: as a profession we are implicitly aware that to understand cities and spatial economics generally we must cope with issues of self-organization and that rather than face what seem to be intractable issues we simply avert our gaze.

Anyway, when we do deal with the question of the organization of space, as urban economists at least must, we generally turn to a class of models pioneered in the early 19th century by von Thünen. Many of you are probably familiar with the von Thünen model, but I want to run through it briefly to make a couple of points about complexity, emergence, and self-organization.

Von Thünen envisaged an isolated town supplied by farmers in the surrounding countryside. He supposed that crops differ in both their yield per acre and their transportation costs, as well as allowing for the possibility that each crop could be produced with different intensities of cultivation. And he asked two questions that might seem to be very different: How should the land around the town be allocated to minimize the combined costs of producing and transporting a given supply of food to the town? How will the land actually be allocated if there is an unplanned competition among farmers and landowners, with each individual acting in his or her perceived self-interest?

We all know the answer to the second question. Competition among the farmers will lead to a gradient of land rents that declines from a maximum at the town to zero at the outermost limit of cultivation. Each farmer will be faced with a tradeoff between land rents and transportation costs; since transportation costs and yields differ among crops, the result will be a pattern of concentric rings of production. In equilibrium the land rent gradient must be such as to induce farmers to grow just enough of each crop to meet the demand, and it turns out that this condition together with the condition that rents be zero for the outermost farmer suffices to fully determine the outcome.

Figure 1.1 illustrates schematically the typical outcome of a von Thünen model. The upper part of the figure shows the equilibrium "bid–rent" curves, the rent that farmers would be willing to pay at any given distance from the town, for three crops. The heavy line, the envelope of the bid–rent curves, defines the rent gradient. Along each of the three segments of that line growers of one of the crops are willing to pay more for land than the others. Thus one gets concentric rings of cultivation, with a quarter section of the layout shown in the bottom half of the figure.

It is worth thinking about this outcome for a moment in terms of the various definitions that have been offered of what complexity theory is all about. If it is really just about complexity, von Thünen models are probably complicated enough to qualify. After all, the problem of which crops to grow where is not that easy: by allocating an acre of land near the city to some one crop, you indirectly affect the costs of delivering all other crops, because you force them to be grown further away. Except in the case where there is no possibility of varying the land intensity of cultivation, it is by no means

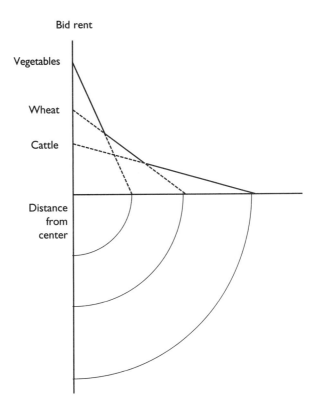

Figure 1.1 The Von Thünen Model. Competition for land around a town leads to the emergence of concentric rings of production

trivial to determine either what should be done or what will happen in an unplanned market.

Suppose that you instead define the subject as Anderson does, as a study of emergence. Then surely the von Thünen model qualifies. At the most obvious level, the concentric ring pattern is hardly something that is immanent in the motives of the farmers. Indeed, the concentric rings will emerge even if no farmer knows what anyone else is growing, so that nobody is aware that the rings are there.

Even more striking is the result that economics has trained us to expect but that remains startling (and implausible) to most noneconomists: the unplanned outcome is efficient, is indeed the same as the optimal plan. More specifically, unplanned competition will allocate crops to land in a way that minimizes the total combined cost of producing and transporting the crops – *not* including the land rent. This is surely as nice an example of emergence as you could want. All farmers are trying to maximize their income and are therefore very much concerned with land rents, yet their collective behavior minimizes a function in which land rents do not appear.

On two of the criteria that have been used to define this enterprise called *complexity*, then, the von Thünen model qualifies. And yet I doubt that many complexity theorists would agree that this was the kind of economics they had in mind. A von Thünen model is not like a hurricane, or an embryo, or a neural network. Why? Because it is not *self*-organizing. The concentric rings of production form around a town whose existence is simply assumed. That does not make it a bad model, but it does make it a limited one. If your question is not simply how land use is determined given a pre-existing town, but rather how land use is determined when the location of the town or towns – indeed, their number and size – are themselves endogenous, the von Thünen model offers no help.

This limitation became painfully clear when the von Thünen model was presented in a new form in the 1960s by Edwin Mills, who substituted commuters for farmers and the central business district for the isolated town, to arrive at his now classic model of a monocentric city.

Again, you are probably familiar with the basic idea if not the details. We imagine a city populated by commuters who want land to live on but must work in a central business district; there may be different types of people who vary in the value they place on the time they spend commuting, the amount of land they want to live on, their willingness to substitute fancy houses for large backyards, and so on. As in the von Thünen model, competition establishes a land rent gradient, which sorts people out into a structure of concentric rings; and as in that model, decentralized location choices lead, through nobody's intention, to an efficient outcome.

Mills's 1967 paper introducing the monocentric city model launched a huge theoretical and empirical literature. And yet in the end that literature proved rather sterile. Part of the problem was purely aesthetic: a model that simply assumes that there is a central business district is deeply unsatisfying if you want to explain cities as opposed to describing them. But urban economists might have swallowed their disappointment if it had not been for another problem: cities do not look like that, and they look less like that with every passing decade.

Remember what I said about greater Los Angeles, with its 16 edge cities far overshadowing its two traditional downtowns. Sophisticated people used to turn up their noses at California cities, with their lack of clearly defined centers – Gertrude Stein declared of Oakland that "there's no there there." She was wrong, of course: there are lots of quite distinct theres there, just no one big there you can call the center. And increasingly that is the way all our cities look. I wrote the draft of this book in Palo Alto, which is part of the San Francisco metropolitan area. San Francisco proper is a compact city, which still epitomizes a certain kind of urbanity. But Palo Alto has little to do with that center. One thinks of oneself as being, not a San Franciscan, but a resident of Silicon Valley; one reads the *San Jose Mercury News* rather than the *San Francisco Chronicle* (it is a better paper anyway), and people who live in Palo Alto are much more likely to commute to the edge city in Sunnyvale than to the vicinity of the Golden Gate. The monocentric city model pictures a metropolitan area as something like a slice from an onion, with rings arrayed around a single center. The reality of all large metropolitan areas in the United States today, even those like New York or Chicago that still have huge, vital downtown office districts, is that they are less like an onion slice and more like Jack Horner's plum pudding, in which edge cities correspond to the plums.

The problem of modeling the structure of the modern polycentric urban area is in large part one of explaining the formation and location of these plums. That is, we cannot avoid the problem of understanding the urban area's self-organization. Part of the process by which that self-organization takes place is, of course, a competition for land that establishes a land rent surface across the urban landscape; in that sense the von Thünen–Mills approach remains essential. But it is at best half the story – and arguably the less interesting half.

CENTRAL PLACE THEORY

Economics as we know it is largely, though not entirely, an Anglo-Saxon tradition. Location theory, however, was long a German tradition, containing at least three streams. One stream follows from the von Thünen analysis of land rent and land use, which we have just discussed. A second stream, associated with Alfred Weber and his followers, focused on the issue of optimal plant location; that literature will play no role in my discussion. But there is a third tradition, which at first sight seems to offer an answer to the problem of spatial self-organization: the central place theory of Christaller and Lösch.

The basic ideas of central place theory seem powerfully intuitive. Imagine a featureless plain, inhabited by an evenly spread population of farmers. Imagine also that there are some activities that serve the farmers but cannot be evenly spread because they are subject to economies of scale – manufacturing, administration, and so on. Then it seems obvious that the tradeoff between scale economies and transportation costs will lead to the emergence of a lattice of "central places," each serving the surrounding farmers.

Less obvious, but still intuitively persuasive once presented, are the refinements introduced by Christaller and Lösch. Christaller argued, and produced evidence in support, that central places form a hierarchy: there are a large number of market towns, every group of market towns is focused on a larger administrative center (which is also a market town), and so on. Lösch pointed out that if a lattice is going to minimize transportation costs for a given density of central places, the market areas must be hexagonal. And thus every textbook on location theory contains a picture of an idealized central place system in which a hierarchy of central places occupies a set of nested hexagons.

The original central place theory story applied to towns serving a rural market. But it is obvious that a similar story can be applied to business districts within a metropolitan area. Small neighborhood shopping districts are scattered across the basins that surround larger districts with more specialized stores, all eventually centering on the downtown with its great department stores and high-end boutiques. Indeed, the hierarchical image is so natural that it is hard to avoid describing things that way.

So why is central place theory not a standard part of the economist's toolkit? Why do the introductory texts find no room for a discussion that uses central place theory to explain why we have business centers, then uses the von Thünen–Mills model to explain the pattern of land use around those centers?

The answer, I believe, is that any economist who thinks hard about central place theory realizes that it does not quite hang together as an economic model. What do we look for in an economic model – or to put it differently, what constitutes an "explanation" from the point of view of economists? I do not think you can sum it up better than Thomas Schelling did in the title of his book, *Micromotives and Macrobehavior*. We feel that we have really managed to shed light on a phenomenon when we show how that phenomenon, the "macrobehavior," emerges (there is one of those words again) from the interaction of decisions by individual families or firms; the most satisfying models are those in which the emergent behavior is most surprising given the "micromotives" of the players. What is therefore deeply disappointing about central place theory is that it gives no account along these lines. Lösch showed that a hexagonal lattice is efficient; he did not show that it would tend to emerge out of any decentralized process. Christaller suggested the plausibility of a hierarchical structure; he gave no account of how individual actions would produce such a hierarchy (or even sustain one once it had been somehow created).

What, then, is central place theory? It is not a causal model. It is probably best to think of it as a classification scheme, a way of organizing our perceptions and our data. Seen in that light, it is a worthy enterprise indeed. After all, in the physical and biological sciences classification schemes have repeatedly served as the basis for great insights – think of the Linnaean classification of species or the periodic table. The point, however, is that classification schemes are only a step on the way: they tell you what, but they do not tell you why. So central place theory is a *description* but not really an explanation of self-organization.

SCHELLING'S SEGREGATION MODEL

The two approaches to the organization of space just described do not quite make it: the von Thünen–Mills approach, which is a model of spatial organization but not of *self*-organization, and central place theory, which is a useful classification scheme but not a causal model. Well, enough of frustration: now let us turn to some approaches that do help explain the self-organization of spatial economies.

I have already expressed my admiration for the work of Thomas Schelling. He is best known for his famous treatise on non-zero-sum games, *The Strategy of Conflict*. But I think that his best book is *Micromotives and Macrobehavior*, an underappreciated classic that had a deep impact on me when I first read it as a wet-behind-the-ears assistant professor. The first chapter of the book is surely the best essay on what economic analysis is about, on the nature of economic reasoning, that has ever been written. And the two chapters on "sorting and mixing" are a wonderful introduction to the idea of self-organization in economics.

If there is any flaw in Schelling's work, it is that he is so clear a thinker that he can often reach deep conclusions with almost no visible technical apparatus and so graceful a writer that he can often make these conclusions seem intuitively obvious. These virtues, I believe, have worked against him. As an amateur anthropologist who has long studied that peculiar culture known as academic economics, it seems to me that an economic idea flourishes best if it is expressed in a rather technical way, even if the technical difficulty is largely spurious.[1] After all, a teacher wants something to do at the blackboard, and a clever student wants something on which to demonstrate his or her cleverness. If a deep idea is conveyed with simple examples and elegant parables, rather than with hard math, it tends to get ignored.

In any case, however, in *Micromotives and Macrobehavior* Schelling presented a simple yet profound model of segregation. The basic idea sounds trivial: segregation results when people prefer not to have too many neighbors who are different from themselves. But Schelling made two much less obvious points. First, mild preferences about the color or culture of your neighbors – preferences that seem on the face of them to be consistent with maintaining an integrated residential pattern – in fact typically lead to a high degree of segregation. Why? Because, even when people have mild preferences of the form "I don't mind having some neighbors of a different color, as long as I'm not too much in the minority," integrated

---

1. You should not, by the way, conclude that I prefer "common sense" to academic discourse. In the real world of affairs, an economic idea is most likely to succeed if it is naively wrong – only then can it appeal to the prejudices of important people.

residential patterns tend to be *unstable* in the face of random perturbations. Second, even if the concerns of individuals are very local – they care about only their immediate neighbors – what emerge are large segregated neighborhoods. Thus Schelling derived, without any fanfare, a theme of many writers on complexity: local, short-range interactions can create large-scale structure.

Characteristically, he made these points with a few simple examples rather than a fully worked-out mathematical model. It is possible to set up such a model, and I am sure that one can even prove theorems about it. I shall reserve my limited mathematical firepower, however, for my own models. For our current purposes, Schelling's approach will be good enough.

Let us, then, imagine a "city" that consists of a number of locations laid out in a square lattice, like a chessboard. For illustration, indeed, let us actually use an 8-by-8 chessboard-size lattice, although the same principles would apply to a much larger domain. And let us suppose that there are two kinds of people – call them black and white, but they could represent any kind of racial or cultural groups, or for that matter different types of businesses that tend to repel each others' customers, like boutiques and auto supply stores.

We now assume that blacks and whites care about the color of their immediate neighbors. (That means the abutting squares on the chessboard.) Their preferences are assumed to be not so much a liking for neighbors the same color as a fear of being isolated. Specifically, Schelling suggested the following rule: an individual with one neighbor will try to move if that neighbor is a different color; one with two neighbors wants at least one of them to be the same color; one with three to five neighbors wants at least two to be his or her color; and one with six to eight neighbors wants at least three of them to be like him or her.[2]

These preferences are consistent with an integrated residential pattern. Consider Figure 1.2, in which # and @ signify the two different groups. Here we have managed to place 60 individuals in a completely integrated pattern, without violating anyone's constraints. That is, complete integration is an equilibrium.

---

2. An equivalent statement of the rule is that each individual requires that at least 37 percent of the neighbors be of his or her own type.

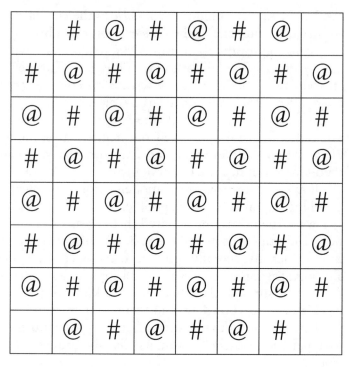

Figure 1.2 **An Integrated City.** Even if people insist that a minimum fraction of their neighbors resemble themselves, it is possible to create an equilibrium residential pattern that is highly integrat-

But are we likely to get a result like this in practice? No, argued Schelling. Although he did not put it quite this way, an equilibrium like the one shown in Figure 1.2 will be *unstable* with respect to some random shuffling and will therefore tend to unravel.

To show this, Schelling took the pattern in Figure 1.2 and messed it up a bit. (His instinct was unerring. My first inclination would have been to start from a completely random allocation of people to locations. It turns out, however, that self-organizing spatial systems yield the greatest order when the initial condition is a small perturbation away from the unstable integrated or flat equilibrium. We will see why in the second part.) Specifically, he extracted 20 individuals at random, both disrupting the pattern and freeing up some room for discontented individuals to move, then disrupted the pattern a bit more by filling five empty squares at random with #s or @s. Figure 1.3 shows the result when I and my random number generator do it.

Figure 1.3   Perturbing the Equilibrium. If the pattern is given some random scrambling, some individuals are no longer content with their location.

It is pretty obvious that in Figure 1.3 some of the people are no longer content with their locations and will move. When they move, however, they will in turn often make those who stay unhappy – either by depriving same-color individuals of neighbors or by shifting the balance against new neighbors of a different color. So a chain reaction begins. To simulate that chain reaction on a computer you would need to specify the order in which people move and how they pick among available locations. If you are doing it by hand, you can be looser about it – it does not matter much. Like Schelling, I did this by hand (using a spreadsheet instead of pennies and dimes on a real chessboard) and just watched the structure evolve. When things settle down, you get Figure 1.4.

Guess what: even though individuals are tolerant enough to accept an integrated pattern, they end up with more or less total segregation. And, even though individuals care about only their

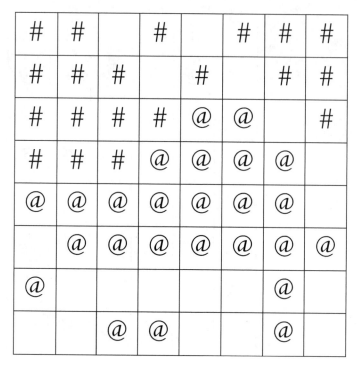

Figure 1.4   A Segregated City. The result is a chain reaction, in which
each move provokes other moves; in the end mild concerns about
being in a local minority produce a completely segregated city.

very local neighborhood, the whole chessboard gets organized into
a white area and a black area.

If # and @ really do represent white and black, it is a nasty outcome.
That, however, is beside the point for this book (except as a reminder
that *organized* does not mean "good"). What I want to emphasize is
that this chessboard city has engaged in a process of self-organization.
Large-scale order – not a nice order, but order nonetheless – has
emerged from a basically disordered initial condition.

This large-scale order emerges because a disordered state, in
which #'s and @'s are evenly mixed, is *unstable*: scramble it a bit and
you start a dynamic process that produces segregation. This is therefore my first illustration of the principle of order from instability –
but do not worry, there will be a lot more.

As you can tell, I love Schelling's model. It is a perfect example of how to wear your sophistication lightly; the way that big conclusions are derived from playing with coins on a chessboard has a special charm. As a model of the self-organization of urban areas, however, it has some limitations. Some of them are problems that I will not try to fix in this book; for example, there is no explicit handling of the market for land, which surely plays a crucial role in mediating the choices of households and firms about where to locate. One problem that I do want to deal with, however, is that Schelling's model is too one-sided: it tells us why birds of a feather flock together but offers no reason why there should ever be more than two flocks.

To see what I mean, ask what would happen if I replaced my 8-by-8 chessboard with a much bigger one, say 1000 by 1000. What would a Schelling-type model predict as an outcome? Well, a huge literature in physics deals with systems not too different from Schelling's model, called *spin-glasses*, and a number of formal results have been derived; but I have to admit that I have not yet taken the time to try to master this literature. Still, a few things seem fairly obvious. The tendency of Schelling's model is always to divide the whole city into two vast # and @ territories. That does not mean that we will necessarily get there: if we start with a random pattern, the chain reaction of moving households will typically die out at some point, leaving the city "frozen" into # and @ domains of varying sizes. (If we set the parameters right, the size distribution will surely obey a power law – but more about that kind of thing later.) We can, however, "melt" these domains if we add a bit of "temperature" to the story, by giving even contented households some small probability of moving just for the hell of it. In that case we will indeed eventually find ourselves with a two-neighborhood city.

Now that is too strong a result. Let us go back to Los Angeles. There are some ways in which the city has a two-zone structure: all the beautiful people live in Beverly Hills, all the nonbeautiful people someplace else. But what is so striking about LA compared with a traditional city is precisely its multipolarity. There is no dominant downtown office district: instead, there are 16 edge cities, spread in a not-too-irregular fashion across the metropolitan area – and, according to Garreau, 8 more in the process of coalescing.

So if we want a story about the self-organization of the urban landscape that reflects the polycentric, plum-pudding metropolitan areas we increasingly inhabit, we need a model that spontaneously produces not only order but some kind of more or less regular repetitive pattern. Let us see if we can construct one.

EDGE CITY DYNAMICS

In a moment I will describe an approach that, it seems to me, sheds considerable light on how office districts – edge cities – may emerge in a polycentric metropolitan area. I believe and hope that the approach will provide a useful metaphor for the emergence of spatial structure in a variety of other contexts. But any experienced economist knows that in presenting a new approach one must begin with some preemptive excuses for the questionable simplifications one has made. So before I even start to describe my framework, let me offer some justifications for the way it is done.

The models that I will describe loosely now and present more fully in the next part will, not surprisingly, represent a kind of idealization of reality. This will not shock anyone: both von Thünen and central place theory begin with the idealization of a homogeneous agricultural plain, which never bore much resemblance to reality and bears even less now than it did when they wrote. What kind of idealization seems legitimate is, however, in the eye of the beholder. My guess is that there are at least four ways in which the approach I am about to present will bother people.

First, there is the issue of geometry. The German tradition in location theory was cheerful about assuming away rivers, roads, and variations in land quality but rigorous about facing up to the consequences of the fact that the Earth's surface is two dimensional. It is common in modern urban economic theory, however, to analyze "long, narrow" cities that are effectively one dimensional (as, for example, in Fujita 1988), a simplification that seems reasonable to me but that would probably have horrified many of the Germans.

Well, it turns out that for my purposes even the one-dimensional city is not quite simple enough, because I want to focus on *self*-organization: the emergence of structure that arises not from inherent differences among locations but from the internal logic of the system. And even in a one-dimensional city locations are distinguished from one another by one crucial aspect: their distance from the ends. That is not simply a formal issue: in reality an office complex near the middle of a metropolitan area will be different from one near the point where subdivisions begin to be mixed with farmland. Because I do not want to deal with that issue, I need to make a silly assumption: my city is not only one dimensional but circular, so there are no ends.[3]

A second issue is the treatment of land and land rents. Modern urban economists are generally willing to play games with geometry, but they tend to be quite fanatical about explicitly modeling land rent – when I presented a model without land rents to one group, a disgruntled urban theorist told me that as far as he was concerned, urban economics was essentially *about* land rent. It is obvious where that attitude comes from: in the von Thünen–Mills model land rent is indeed the crucial ingredient. But I am asking a different question and will ask you to bear with an approach that recognizes land scarcity at best in an implicit, reduced-form way.

A third issue is the treatment of expectations. I will be telling a dynamic story; it is a story about how edge cities *evolve*. But I do not want to worry about forward-looking behavior. To many economists, raised in an environment of rational expectations theory, that by itself disqualifies a model. Most of you probably do not care, but for those of you who do, I urge you to put your prejudices on hold.

Finally, I have a personal rule that I am about to break. In general, I do not like simply assuming the existence of external economies and diseconomies. It often seems too close to assuming your conclusions. (I know of one economist who tried to explain his work to a group of physicists, one of whom sarcastically said, "So what you're telling us is that firms agglomerate because of agglomeration economies.") Most of my work on economic geography has focused on trying to *derive* external economies out of the interactions among scale economies, transportation costs, and factor mobility – to make external economies an emergent property. I actually arrived at the approach I am getting to in the context of such a model, one that I shall present in the second part. But it turns out that it is easiest to explain the basic idea simply by assuming external economies, and so I will temporarily suspend my rule.

-----

3. Alternatively, one can imagine a city of infinite extent. Los Angeles may approximate this condition.

Now that the excuses are out of the way, let us get to the substance.

Imagine a metropolitan area, which we can think of as a homogeneous expanse of identical housing developments – except that, rather oddly, the population is distributed not in a two-dimensional sprawl but in a narrow ring, and travel is possible only along that ring's circumference. And suppose that there is some business activity – it could be office work, it could be retailing – that depends on the spread-out population both as a market and as a source of labor.

Let us also, unobjectionably, suppose that the decisions by businesses about where to locate are interdependent. That is, the desirability of any one site as a business location depends on where all of the other businesses are located. And we may also safely suppose that businesses migrate over time from less to more desirable sites.

Clearly the dynamics of this process depend on the nature of this interdependence. One can imagine two general sorts of interdependence. On the one hand, businesses might dislike having other businesses nearby, because they compete for customers, workers, or land. Call these considerations *centrifugal* forces, forces that promote dispersion of business. On the other hand, businesses might like to have other businesses close, because they attract customers to the area or help support a greater variety of local services.[4] Call these *centripetal* forces, forces that tend to make businesses clump together. If only centrifugal forces existed, businesses would spread themselves evenly across the landscape. If there were only centripetal forces, they would rush together into one big clump.

But what explains the polycentric, plum-pudding pattern of the modern metropolis? In general, a model that would explain this pattern must meet two criteria:

Criterion 1. There must be a tension between centripetal and centrifugal forces, with neither too strong.

Criterion 2. The range of the centripetal forces must be *shorter* than that of the centrifugal forces: businesses must like to have other businesses nearby, but dislike having them a little way away. (A specialty store likes it when other stores move into its

---

4. Joel Garreau writes: "Five million square feet is a point of spontaneous combustion. It turns out to be exactly enough to support the building of a luxury hotel. It causes secondary explosions; businesses begin to flock to the location to serve the businesses already there."

shopping mall, because they pull in more potential customers; it does not like it when stores move into a rival mall 10 miles away.)

And that's all that we need. In any model meeting these criteria, any initial distribution of business across the landscape, no matter how even (or random), will spontaneously organize itself into a pattern with multiple, clearly separated business centers.

Does this proposition sound obvious? Maybe so (although I did not have it clear in my own mind until after I started playing with mathematical models). But there is more. For a wide variety of specific models in which criteria 1 and 2 are met, any initial distribution of business across the landscape will evolve not merely into a pattern with several business centers but into a pattern in which these centers are roughly evenly spaced, with a characteristic distance that depends on the details and parameters of the model but not on the initial distribution. And the smoother is the initial spatial distribution of the businesses, the more even their eventual spacing.

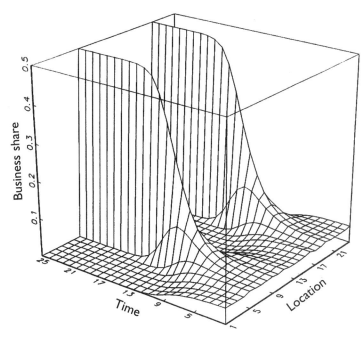

Figure 1.5   **The Evolution of Edge Cities.** An ititially almost uniform distribution of business accross the landscape evolves spontaneously into a highly structured metropolis with two concentrated business districts.

At this point I had better show you an illustration. It turns out, not too surprisingly, that computer simulation – and computer graphics – are an invaluable aid to thinking about self-organizing systems. Figure 1.5 is a sample run of a model of a "city" consisting of 24 locations around a circle. The locations are shown on the X axis; bear in mind that location 24 is next to location 1. I started the run with a fairly but not perfectly even allocation of business across these locations,[5] then let it evolve according to a rule that caused businesses to move toward locations that were highly desirable. "Desirability" of a location was both positively and negatively affected by the number of businesses at other locations, with both effects declining with distance but with the positive effects declining faster than the negative effects. The Y axis of the figure shows the passage of time; the Z axis shows the share of the businesses in each location at each point in time. Thus the figure gives you a sort of frozen portrait of the whole simulated history.

Let us look a bit at Figure 1.5; the picture contains quite a lot of information. The right edge of the calculated surface, which represents the initial geographical distribution of business, is almost a horizontal line. That is, I have started my "city" off with almost no spatial organization. But eventually the surface rears up into a pair of dorsal fins:[6] all of the businesses end up in locations 8 and 20. The city has spontaneously developed a strong spatial structure.

This may not be surprising. But now notice which locations get the businesses. In a 24-location circular city, locations 8 and 20 are exactly opposite one another. That is not an artifact of the particular starting position: if you run the model repeatedly with these parameters, but with a different initial spatial distribution of business each time, you will consistently get two business concentrations opposite each other.

By the way, I do not want you to get the impression that there is something special about cities with two business districts. Figure 1.6 shows a typical run of the same model with somewhat different

---

5. Each location received a weight $s + u_i$, where u_i was a location-specific random variable between 0 and 1, and s was a "smoothing" parameter, set equal to 5 for this run. Then each location was assigned a share of the businesses proportional to its weight.

6. Robert Gordon of Northwestern University has dubbed Figure 1.5 the '59 Cadillac model.

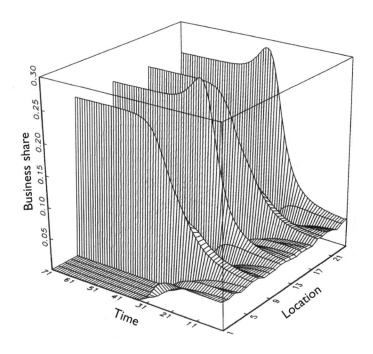

Figure 1.6   Same Story, Different Parameters. With somewhat different parameters, the metropolis evolves four business districts.

parameters. In this case we get *four* business districts, equally spaced around the circle. The general principle, then, is that the city organizes itself into a structure with a characteristic distance between business districts; if there are only two, this implies that they face each other.

Let us go back to Figure 1.5 and look at one more thing. Look at how the almost flat initial surface evolves over time. Although in the end only two widely separated locations end up with business concentrations, in the early stages of self-organization it does not look as if the winning locations are growing at the expense of their neighbors. On the contrary, initially not only the winning locations but those nearby grow. The surface seems to undulate, with waves rising up out of the plain. Only after some time has passed do these waves gather themselves into dorsal fins, with the eventual centers cannibalizing their neighbors. Again, this impression is not unique to a particular run: it is a consistent feature, whatever the initial distribution.

It turns out that this image of an undulating surface, with the waves growing over time, is the key to understanding the process of self-organization. To fully explain why will take a little time, and I'll reserve that for the second part (and the technical appendix). But let me offer a preliminary view.

Remember that I started the simulation in Figure 1.5 with an almost but not exactly flat distribution of businesses around the circle. The deviation of that distribution from perfect flatness may be represented as an irregular, wobbly line. But it is generally true that, to use technical language, an irregular wobble can be thought of as the sum of many regular wobbles at different frequencies. In particular, an irregular wobble around a circle can be expressed as the sum of one line that wobbles once as it goes around the circle, another that wobbles twice, a third that wobbles three times, and so on. (Some of you know that I am talking about a Fourier series.)

Now here is the point: in any model that satisfies my two criteria, some of these component wobbles – wobbles that go certain particular numbers of times around the circle – will tend to grow over time (although wobbles at other frequencies[7] may tend to die out). And, because the decomposition of an irregular wobble will ordinarily contain wobbles of all possible frequencies, an even distribution of business around the circle is *unstable*. Some of the component wobbles will grow, creating an increasingly uneven spatial distribution of business. Order from instability!

Why, however, is the eventual spacing of business so regular? Because wobbles that go different numbers of times around the circle will grow at different rates. If the initial distribution of business is sufficiently smooth, after a while the deviation from smoothness will be dominated by whichever frequency wobble grows the fastest – by the *most unstable* wobble. And the frequency of this wobble, the number of times it goes around the circle, will determine where the peak business concentrations are located. The undulations you see in Figures 1.5 and 1.6 are that most unstable wobble taking over the distribution of businesses. For the parameters used to generate Figure 1.5, the most unstable wobble goes around the circle twice; for those used to generate Figure 1.6, the most unstable wobble goes around the circle four times. Not only does instability create

---

7. Remember that these are spatial frequencies: these are wobbles in space, not time.

order, the form of that order is dictated by a sort of principle of maximum instability.

What about the last part of the figures, where the wobbles gather themselves up into spikes? Well, the formal answer is that everything that I have said is valid only for a linear approximation to the model, which breaks down when the spatial distribution of business gets too uneven. Less formally, in the early stages of self-organization the most favored locations can grow by pulling businesses away from distant locations. Once there are no more businesses in the large gaps between business centers, they can continue to grow only by eating their neighbors. The important point, however, is that the locations of the winners are determined in the earlier stage: the distance between edge cities is determined by the wavelength of the most unstable wobble.

OK, let us stop and take a deep breath. Even though I have avoided any formal modeling, you may at this point have suddenly realized how seemingly abstract, how unrelated to the details of freeways and shopping malls, skyscraper construction and fast food consumption, this has gotten. That is very much the style of complexity theorists: indeed, the whole rationale of the field is the idea that common principles may apply to subjects with very different details. Still, has the abstraction led us into a story that conflicts badly with reality?

My guess is that many readers will object to the implied regularity of the result – those equal-size, regularly spaced business concentrations. In reality edge cities are not all the same size or equally spaced across the landscape. That does not worry me, however. After all, the real landscape is not homogeneous. There are highways, whose intersections make particularly favorable sites for business, variations in the pleasantness or buildability of sites, and for that matter real metropolitan areas do have centers and edges and thus are prima facie not undifferentiated. Furthermore, the result that business districts are evenly spaced is true only for an imaginary history in which business starts out spread almost evenly across locations; because the real histories are not like that, we should expect a more irregular result.

But if real cities do not evolve in the way I have just described, does that mean that the whole approach is irrelevant? I think not, but I need to introduce some more concepts and examples to explain why.

2

Complex Landscapes

In his wonderful book *Nature's Metropolis: Chicago and the Great West*, the historian William Cronon distinguishes between two landscapes in which urban evolution takes place. One is the natural landscape: the mountain ranges, rivers, and lakes that are givens of the environment. The second is the created landscape of railroad lines, canals, farming patterns, and cities themselves that results from human decisions. Cronon argues that in the modern world the created landscape, which he refers to as *second nature*, has become far more important as a determinant of location than the "first nature" in which it is embedded: Chicago's role as a Great Lakes port was quickly overshadowed by its role as a rail hub. And second nature is often self-reinforcing: railroads aimed at Chicago because it was the economic center of its region and thereby made its centrality all the greater. That is, the landscape of second nature is inherently dynamic.

But how does one visualize a landscape that changes over time? One answer is a historical atlas: a sequence of maps that traces out the changes. In some cases it may be possible to do a bit better by drawing a kind of temporal relief map in which the time dimension, and thus the evolution of the landscape, is more or less imperfectly represented. Indeed, Figures 1.5 and 1.6 are just that: they show how some imaginary landscapes change over the course of their imaginary histories. But even when you can do this, all that you get is a description of what happened; *why* it happened is at best implicit.

So what do you do? Anyone who has worked on formal models of dynamic systems knows the answer: you try to draw a picture of yet a third landscape, a "phase space" in which each point summarizes the position of the system at a point in time and in which the rules that govern the system's evolution are translated into "laws of motion" in that abstract landscape.

Phase space representations of dynamic systems are extremely common in modern economic analysis. In general, however, we tend to focus on only a narrow range of possible types of landscapes – indeed, basically on only two fairly simple forms. Before the mid-1970s nearly all dynamic models in economics were globally stable; that is, the phase landscape was assumed to be like a bowl, a single basin of attraction in which all points drain to a single long-run equilibrium. Since about 1975 it has become common also to work with models in which the phase landscape looks like a saddle – and in which some set of forward-looking variables, such as asset prices, is determined by the assumption that the economy is always on the ridge that is the only path to long-run equilibrium.[1]

The literature on complexity, however, is largely concerned with systems in which the dynamic landscape looks like neither a bowl nor a saddle; indeed, it is often concerned with "rugged landscapes" (as Stuart Kauffman puts it) that look like the South Dakota badlands.

What aspects of a dynamic system lead to a rugged phase landscape? I have no general answer, and I do not know if there is one. Many of the models in the complexity literature, however, have a similar setup.[2] The modeler represents his or her system as an ensemble of many components, each of which is at any particular time in one of several states: magnetic dipoles that are oriented up or down, neurons that are firing or quiescent, genes that have one character or the other, and so forth. And future changes in these

---

1. Strictly speaking, one can represent a two-dimensional dynamic system as a three-dimensional relief map only if the system acts as if it were following the gradient of some potential function. This is sometimes reasonable, but not always: the evolution of a single species is in effect maximizing something we can call fitness, but the coevolution of predators and prey is not. Nonetheless, the language commonly used to describe phase space, with its basins of attraction, saddle paths, and so on, draws heavily on the relief map metaphor; and I at least view the poetry as worth the potential confusion.

2. See, for example, the influential paper by Hopfield (1982) and compare it with the "NK model" of Kauffman (1993).

states are linked in some way: the energy of a spin-glass, and thus its likely direction of change, depends on whether nearby dipoles are oriented in the same direction or not; neurons excite or inhibit each other; the effect of changing an individual gene on an organism's fitness, and thus the likelihood that a mutation will survive, depends on what other genes it has. Such a system, it turns out, will produce a complex dynamic landscape as long as two conditions are met: the responses of the individual units must be discrete (neurons either firing or not), and there must be a mixture of positive and negative feedback (neurons both excite and inhibit each other).

Now think about spatial economies. They are systems in which many components (firms) are at any particular time in particular states (locations) and in which changes in these states are linked (through agglomeration economies and diseconomies). If there are significant economies of scale, firms will choose only a few discrete locations; and there will usually be a mixture of positive and negative feedback between these choices. Therefore we might well expect spatial economies to have complex, rugged dynamic landscapes.

And indeed that is what you find when you look at even quite simple models of spatial economies. Over the last few years, I have spent a lot of time trying to understand the behavior of a model almost as minimalist as the edge city model I introduced a little while ago. (In fact, as I shall explain in Part II, the two models behave in very similar ways.) In this structure we imagine that there are two factors of production: immobile agricultural workers and mobile manufacturing workers. Manufacturing is a monopolistically competitive sector characterized by both increasing returns at the level of the firm and transport costs. The interaction among factor mobility, increasing returns, and transport costs generates forces for agglomeration: firms tend to concentrate production in locations with good access to markets, but access to markets is good precisely where other firms are concentrated. Working against these "centripetal" tendencies, however, is the "centrifugal" pull provided by the geographically dispersed agricultural sector.

If we add some rudimentary laws of motion, say the assumption that manufacturing workers tend to move to locations that offer relatively high real wages, we get a dynamic story in a phase space defined by the allocation of manufacturing workers across locations. And this dynamic landscape can easily be very complex.[3]

An easily shown example is the three-region case. Suppose that there are three equidistant locations, with equal agricultural labor forces. The allocation of the manufacturing work force among the three locations can be represented as a point on the "unit simplex": a triangle whose corners are at the points (1, 0, 0), (0, 1, 0), and (0, 0, 1) in a space whose axes are the share of manufacturing in each location but that can conveniently be pasted onto a two-dimensional page. Starting at any given point on that simplex, one can let the model evolve and see where it ends up. To draw the picture analytically is extremely difficult, but it is straightforward to compute numerical examples.

Figure 2.1 shows what I get for the most interesting range of parameters. (There are only three parameters in the model: the elasticity of substitution among products in the manufacturing sector, set for this example at 4; the share of manufactures in expenditure, set at 0.2; and the transport cost between any two locations, set at 0.4.) At each of a number of points on the simplex, representing an initial allocation of manufacturing workers, I draw an arrow representing the direction and speed of "flow". It turns out that there are four equilibria: three in which all manufacturing is concentrated in one location, one in which there is an equal distribution of manufacturing across the locations. There are correspondingly four basins of attraction: a central basin that leads to the equal division outcome, and three flanking basins that lead to concentration.

The landscape can become far more complex when there are more locations. Suppose, for example, that we consider an example in which there are 12 locations, laid out in a circle like the numbers on a clock. (Twelve is a particularly convenient number because it is a fairly small number with a large number of divisors.) The dynamics once again take place on a unit simplex – but this time an 11-dimensional one. This is hard for most of us to visualize. We can, however, get a good idea of the properties of the model exper-

---

3. The kind of complex landscape that can arise in models of economic geography can, of course, arise in many other economic contexts as well. Most notably, the choice among several technologies subject to network externalities will present a very similar picture, with the landscape complex if none of the technologies has too strong an inherent advantage and the externalities are sufficiently powerful.

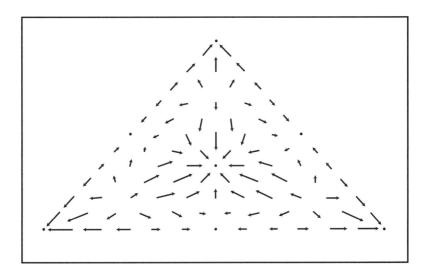

Figure 2.1 Basins of Attraction. A three-region economic model can end
up with four different locational patterns, depending on initial
conditions.

imentally, by starting with a number of random allocations of manufacturing across locations and seeing how they evolve.

Figure 2.2 illustrates a typical run. The initial random allocation
of manufacturing eventually organizes itself into two manufacturing
concentrations, at locations 6 and 11; that is, 5 apart. In the course
of a number of runs with these parameter values, I got two concentrations 5 apart about 60 percent of the time, two concentrations 6
apart on almost all other occasions. At rare intervals a run would
lead to three equally spaced concentrations.

On a circle with 12 locations, there are 12 ways to place two
markers 5 apart, 6 ways to place them 6 apart, and 4 ways to place 3
equidistant markers. So it appears that with these parameters the
model implies a landscape with 22 basins of attraction – rugged ter-
rain indeed. In such a world the location of economic activity, and
to some extent even the structure of the resulting economic geography, would depend crucially on initial conditions, which is to say
on historical contingency.

Now, finally, we get back to the question that led us off on this
side trip through phase space. The beautifully simple, aesthetically
pleasing "'59 Cadillac" pictures in Figures 1.5 and 1.6 were based

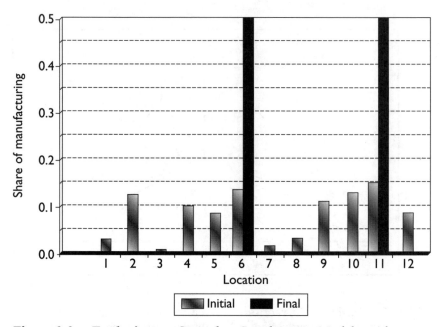

Figure 2.2. **Exploring a Complex Landscape.** Models with more regions can be explored via simulations; here is a typical run for a 12-region example.

on the assumption that the spatial economy started with an almost uniform distribution of business across space. But that is not how real economic history works — and we have just seen that spatial economies typically must have complex dynamic landscapes in which where you end up depends a lot on where you start. So is the simple picture basically irrelevant?

I think not — because although spatial models easily generate complex dynamics, lurking within that complexity we often find surprising simplicity.

THE EMERGENCE OF ORDER

The most provocative claim of the prophets of complexity is that complex systems often exhibit spontaneous properties of self-organization, in at least two senses: starting from disordered initial conditions they tend to move to highly ordered behavior, and at least in a statistical sense this behavior exhibits surprisingly simple regularities: for example, a power law distribution relating the sizes and frequencies of earthquakes.

Let us look again at what I said about the 12-region model of economic geography whose outcome is represented in Figure 2.2. The model appears to have a dynamic landscape with 22 basins of attraction. And yet, looked at a different way, the results are surprisingly ordered. Starting with a random allocation of manufacturing across space, the model always organizes itself into a highly ordered structure in which manufacturing is concentrated in two or three equal-size concentrations. Furthermore, although there are many such equilibrium structures, they share strong similarities: all involve roughly equidistant city locations, with a fairly narrow range of typical distances between cities.

What we saw in the edge city model was these characteristic features of equilibrium structures in an extreme form: if we started that model from an almost uniform initial distribution of business (corresponding to a starting point near the center of the simplex in Figure 2.1), we got business concentrations exactly evenly spaced around the circle, with an invariant distance between concentrations. The same occurs in this model: if I take the parameters used to generate Figure 2.2 but restrict myself to initial positions in which the distribution of manufacturing is sufficiently flat, I will consistently get a picture like Figure 1.5 – the economy organizes itself into a structure with two cities, exactly opposite one another.

What this suggests to me is that the fake history, in which a highly regular spatial structure emerges from an almost unstructured initial position, can be viewed as a sort of model of the model. That is, the precise regularities of that special case help us understand the rough regularities of the more general case.

Nor do I think that this insight is merely about modeling. It seems reasonable to speculate that the immensely more complex landscape that determines the real geography of the world economy has its own underlying approximate simplicities. That is, there may be many possible outcomes, depending on initial conditions – Silicon Valley might, given a slightly different sequence of events, have been in Los Angeles, Massachusetts, or even Oxfordshire. But some broader features may be more or less independent of historical contingency.

And in fact one regularity in spatial economics is so spectacular in its exactness and universality that it is positively spooky. That regularity involves the size distribution of cities, and it leads us to our other principle: order from random growth.

3

An Urban Mystery

Let us start with a picture. Suppose that you take all of the metropolitan areas listed in the *Statistical Abstract of the United States*, 130 in number, and rank them by population. And suppose that you plot the ranks of these areas against their populations, using a logarithmic scale. What you get is Figure 3.1.

What is interesting about this figure? Of course the line relating rank to population is, by definition, downward sloping. But there are two things about that line that did not have to be true. First, it is pretty close to a *straight* line: there seems to be something close to a log–linear relationship between rank and population. That in itself is pretty interesting; surely it suggests that some hidden principle is at work. But even more interesting is the slope of that approximately linear relationship: it is very close to –1.

There is another way of putting these results, which is also another name for what is sometimes called *Zipf's law*: it is the *rank–size rule*. This rule says that the population of a city is inversely proportional to its rank. If the rule held exactly, the number 2 city in a country would have half the population of the biggest city, the number 3 city one-third that population, and so on. Obviously the rule does not hold exactly. If you look at only the largest metropolitan areas, you may be unconvinced: Los Angeles is considerably more than half as populous as New York. But once you get down the ranking a bit, the fit starts to become almost terrifyingly exact. For example, the 10th ranked metropolitan area in the United States is Houston, with 3.85 million people. The 100th ranked area is Spokane, Washington, with 370,000 people – close enough to

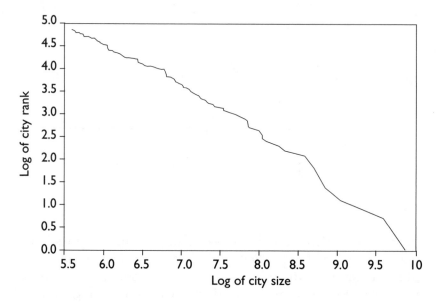

Figure 3.1 Zipf's Law: The size distribution of metropolitan areas in the
United States is startlingly well described by a power law, with an
exponent very close to 1.

Source: *Statistical Abstract of the United States,* 1993.

one-tenth of Houston to be well within reasonable uncertainty of
definition and measurement. If you regress the log of rank on the
log of population, you get a coefficient of –1.003, with a standard
error of only 0.01 – a slope very close to 1 and very tightly fitted.

We are unused to seeing regularities this exact in economics – it
is so exact that I find it spooky. The picture gets even spookier
when you find out that the relationship is not something new –
indeed, the rank–size rule seems to have applied to U.S. cities at
least since 1890! Figure 3.2 shows data taken from *Historical Statistics
of the United States,* which reports the number of "urban places" in
specified size ranges; I show the number with more than 100,000,
more than 250,000, more than 500,000, and more than 1 million
for 1890, 1940, and 1990. (These data are not quite comparable to
those in Figure 3.1, because of the difference in definition between
an urban place and a metropolitan area.) The picture is not perfect,
but if you had any reason to believe that there was a fundamental
reason to expect a straight line with a slope of –1, you would find
this picture compelling evidence in favor of your theory!

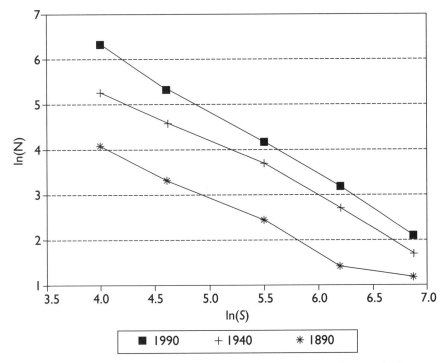

Figure 3.2 Zipf's Law over Time. The same power law has worked reasonably well for at least a century.

Source: *Historical Statistics of the United States.*

Zipf's law is not quite as neat in other countries as it is in the United States, but it still seems to hold in most places, if you make one modification: many countries, for example, France and the United Kingdom, have a single "primate city" that is much larger than a line drawn through the distribution of other cities would lead you to expect. These primate cities are typically political capitals; it is easy to imagine that they are essentially different creatures from the rest of the urban sample.

What could explain the existence of something that looks suspiciously like a universal law on city sizes? Bear in mind that there are two things we need to explain. We need to understand why the rank–size relationship is nearly linear in the logs, and we need to understand why it stays so oddly close to a slope of –1.

When urban economists attempt to explain the rank–size rule, they usually argue that it reflects some sort of hierarchy of central places. Here is how this reasoning goes. First, we point out that an alternative way to express the linearity of the rank–size relationship is to say that the distribution of city sizes follows a *power law* – the number of cities whose population exceeds some size S is proportional to S to some negative power.

Next, we note that a power law distribution is consistent with a simple hierarchy. Suppose, for example, that we think of a managerial hierarchy, in which each manager has the same span of control; say, each manager oversees 10 subordinates. And suppose also that, say, three of those subordinates, plus their subordinates (if any) must work in the same location as their superior. Then there will be a hierarchy of managerial sites: for every 7 individual field offices, there will be a central office with 4 people; for every 7 such offices, there will be a more central office with 13 people; for every 7 offices there will be yet another level of offices with 40 people each; and so on. If you can imagine this hierarchy going on forever, you will find that each successive type of office will have approximately three times as many employees as the previous level. Because each office supervises seven off-site offices at the previous level, this means that the distribution of office sizes will be a power law with an exponent of –7/3, or –2.33.

Now imagine that the sizes of central places reflect just such a hierarchy: each local central place holds field offices, each member of the next level contains subregional offices, and so on – you may have a story about the apparent power law on city sizes.

Stories of this kind have come in for some unjustified criticism. Taken literally, they seem to imply a set of discrete city sizes, not the smooth distribution of Figure 3.1. The usual answer, which is surely right, is that we need not suppose that the relationships are exact and that with a little noise the distribution could well appear smooth.

And yet I am convinced that such hierarchical explanations of the rank–size rule are wrong, for three quite different reasons.

First, the whole image of central place theory, in which towns and cities form a structure whose spacing is dictated by a dispersed agricultural population, seems hard to credit in a modern economy where land-based activities like farming play such a small role. Indeed, a glance at the nighttime satellite photo of the United States that I keep on my wall shows a picture that looks nothing like a regular central place pattern. Instead, we see great belts of light in a few areas, vast stretches of near-darkness in between.

Second, the hierarchical story depends on the constancy of a number of parameters that we have no particular reason to think are constant. Why should the span of control of managers be the same when they are supervising very different levels of activity? Why should the number of people in an office at each level be a constant multiple of the number at the level below? The assumptions seem a bit too close to the conclusions.

And finally, the story fails to explain the full, astonishing fact of the rank–size rule. A hierarchical model suggests why a hierarchy of cities might more or less obey a power law on sizes; it does not explain why they should almost exactly follow not only a power law, but the very specific law that size is precisely inversely proportional to rank. Indeed, the example I have just given has size falling much too slowly with rank to be consistent with the data.

But, if we abandon the idea of a central place hierarchy, what possible explanation can we offer for Zipf's law?

METEORITES, EARTHQUAKES, AND CITIES

The mysterious regularity implied by Zipf's law is not unique to urban economics. On the contrary, power law regularities appear in many other fields. The frequency with which meteorites exceeding any given diameter strike the Earth follows a power law. There is a famous log–linear relationship, the Gutenberg–Richter law, between the sizes and frequencies of earthquakes. Plot the number of animal species that exceed a given size and you will again find the same kind of relationship. So we are not alone in our mystery.

Where do such strangely simple relationships come from? Loosely speaking – and much of the reasoning in this area is surprisingly loose – it seems that you get results like this when three criteria are satisfied. First, the objects you are studying are subject to substantial *growth* over time. Second, the growth rate of any individual object is *random*, so that over time you get a wide range of different sizes. Finally, however – and crucially – the expected rate of growth (though not necessarily its variance) must be *independent of scale*: large objects must grow on average neither faster nor slower than small ones.

Take the case of meteorites. The current best-guess model starts with a bunch of fairly big rocks out there in space. Every once in a while, these rocks collide and break into smaller fragments. (This is a growth process, in which the expected growth rate is negative.) Because collisions and fragmentation are random, you get a wide variety of meteorite sizes; because big rocks and small rocks are just as prone to fragmentation, you get a power law.

Earthquakes are a more subtle and disputed case. The Santa Fe theorist Per Bak has proposed a model in which earthquakes are produced by a kind of chain reaction. Think of the land near a fault line as consisting of many blocks of earth, each under stress countered by friction. If the stress exceeds a critical level, a block will slip – and in so doing transfer stress to its neighbors, some of which may in turn be triggered into slipping, and which may trigger their neighbors in turn, and so on. Bak has shown that if the average level of stress on the blocks is close to a critical level, such chain reactions constitute a sort of growth process with the right features to produce a power law, random growth that is independent of scale. And he argues that earthquake zones tend naturally to evolve precisely to that critical level of stress, in what he calls *self-organized criticality*.

Can we tell a story about cities that is anything like these cosmic accounts? Indeed we can. In fact, Herbert Simon proposed such a story more than 40 years ago, in one of those papers that should have been extremely influential but that somehow, whether because they did not fit the zeitgeist or because they were written in the wrong style, were largely ignored. Perhaps this book will suffer the same fate, but anyway let me offer my own version of Simon's model.

SIMON ON ZIPF

I would describe Simon's story (I call it a story, because it is almost too nihilistic to call a model) as one of "lumps and clumps." We imagine an urban system with a growing population. Population does not, however, arrive smoothly; it arrives in discrete "lumps" that are neither too small nor too big. I shall explain what that means in the next part.

Where does a lump land? With some probability π the lump lands in a new place; that is, a new city is formed. With a probability $1 - \pi$ the lump attaches itself to an existing "clump," which consists of one or more lumps. The probability with which any given clump attracts a new lump is proportional to its population.

And that is it. If you like, you can easily give the process a minimalist economic interpretation. Think of a "lump" as being the number of people employed directly or indirectly by an entrepreneur. Imagine that new entrepreneurs are usually people inspired by (perhaps working for) existing enterprises; so any given city will generate new ideas in proportion to its population. And finally imagine that entrepreneurs usually, though not always, stay near their point of original inspiration. Then you have Simon's story, more or less.

Now this story feels all wrong. After all, it seems to imply that the size of a city is essentially irrelevant! In the story, entrepreneurs "give birth" to additional entrepreneurs at a rate that is independent of how big the city is. Surely we believe that there are all sort of systematic differences among cities of different sizes – big cities are more creative but more costly to live in, small cities typically get the industries that have matured in the dynamic big-city environment, and so on and so forth. This nihilistic story *must* be too simple to capture the complexities of reality. Except that the reality of the city-size distribution is *not* complex – it is startlingly, spookily simple. And Simon's story does explain that simplicity.

Consider Figure 3.3. It represents the result of a run of a Simon story with π set equal to 0.2; I started with 10 "seeds," initial cities

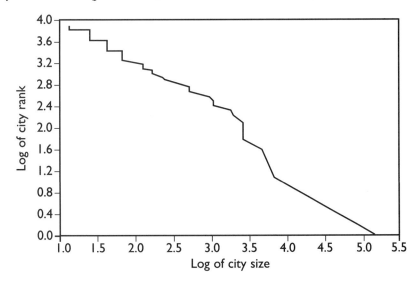

Figure 3.3 Lumps and Clumps. A model of random urban growth produces results that are similar to the actual data – but not as neat!

each consisting of one lump, then allowed the population to increase by a factor of 100. The figure shows the rank–size distribution for the top 50 cities. It is not quite as neat as the real data – it is, indeed, a peculiar feature of the whole Zipf issue that the real data are cleaner than anyone's model! But it does bear a strong resemblance to the real distribution, both in that it roughly obeys a power law and that the slope of the relationship is close to -1.

It turns out that Simon's process always does, if you run it long enough, generate something that looks like a power law distribution for the sizes of larger clumps. The exponent on this distribution depends on the probability of forming new cities, π; numerical experiments suggest that you get an exponent close to 1 when π is small. In fact, Simon showed – in a completely impenetrable exposition! – that the exponent of the power law distribution should be $1/(1 - \pi)$. I shall offer some intuition on this in Part II, including what I hope is a more user-friendly derivation than the one Simon provided. The point, however, is that, if we accept the simplistic justification for Simon's process just given, there is a natural economic reason not only for the existence of a power law but for this power law to have a slope near 1: what the data are telling us is that only a small fraction of entrepreneurs move far from the source of their ideas.

4

Principles of Self-Organization

The more mystical prophets of "complexity" express the hope that we will eventually arrive at universal laws of self-organization that apply to all complex dynamic systems. A more modest goal would be to identify families of situations, possibly in very different contexts, in which similar principles and behavior arise; it would be disappointing but not crushing if it turned out that no common principle links the different families and if the families do not cover all possible situations.

At least at this point I remain in the camp of modest objectives. I want to make no grandiose claims about having found a theory of everything. And yet the principles I have identified apply to a wide class of phenomena in a startlingly wide array of fields.

ORDER FROM INSTABILITY

Try filling a pail with a mixture of golf balls and Ping-Pong balls. If you are careful, you can manage to layer them together so that Ping-Pong balls are as likely to be on the bottom as on the top, golf balls as likely to be on the top as on the bottom. But shake the pail, and pretty soon it will be all Ping-Pong above, golf below. The initially disordered structure organizes itself, because a disordered structure is *unstable* when subject to random shocks.

The family resemblance of the Ping-Pong/golf example to Schelling's urban segregation model should be obvious. Indeed, models in which order emerges from instability are pervasive in the physical sciences. The classic illustration is convection. Heat the bottom of a pan of water and cool its top. Once the temperature gradient reaches a critical level, a motionless equilibrium becomes unstable. The water then organizes itself into a pattern of hexagonal convection cells.

Meteorology, which is concerned largely with convection in the wild, is often thought of as the home of chaos. But it is also a subject preoccupied with the emergence of large-scale if temporary order. In his fascinating book on the subject, Edward Lorenz, who "discovered" chaos, nominates as the greatest meteorological theorist of all time the Norwegian Vilhelm Bjerknes. What was Bjerknes's big insight? That "the reason that vortices and other structures of continental size must be present in the atmosphere is not the dynamic impossibility of a flow pattern without them . . . it is the instability of such a pattern with respect to inevitable disturbances of large horizontal extent but small amplitude."

Perhaps most surprising, the theme of order from instability appears in biology as well. One of the biggest questions in biology is morphogenesis: How do we get from a single fertilized egg to an organism that not only requires many different kinds of specialized cells but particular types of cells in particular places? An individual fruit fly cell presumably does not think to itself, "I am part of a wing," yet cells collectively seem in effect to decide to become different parts of the organism. Experiments suggest that cells indeed behave as if they knew their own polar coordinates!

It turns out that there is a classic 1952 theoretical paper on morphogenesis, which may explain this remarkable fact, by none other than Alan Turing, of computer fame. Turing pointed out that chemical systems – for example, a two-chemical system in which one chemical catalyzes both its own production and the production of the other, the second inhibits the production of the first, and the two diffuse at different rates – can exhibit spontaneous spatial self-organization. Small fluctuations in the initial distribution of the chemicals will tend to grow over time; in particular, any such system will have a particular "preferred wavelength," a particular size of fluctuation that tends to grow fastest and thus to dominate the emergent pattern. He then argued that, if one of these chemicals is a "morphogen," a substance that tells a cell what to become, this self-organizing chemical system can allow cells to behave as if they know where they are in relation to one another.

Does this sound familiar? The edge city model described previously was inspired by Turing's reaction–diffusion model. It differs greatly in the details: in particular, in the biological analysis it is essential to model cells as being affected directly only by their immediate neighborhood, whereas in an urban model nothing is disturbing about the idea that a mall's profitability is affected by conditions 10 miles away. Still, there is a clear family similarity, and some essential elements of the mathematical formalism, which I shall discuss in the next part, can be carried over directly. So as I promised at the beginning, a city is indeed something like an embryo.

I believe that the principle of order from instability, and in particular the notion that self-organizing systems tend naturally to be dominated by the most unstable fluctuations, is a powerful idea. It has certainly helped me to think about metropolitan structure – or perhaps we should just call it *urban morphogenesis*. I suspect that it will often be a way to cut through the complexity of what may seem hopelessly difficult models.

ORDER FROM RANDOM GROWTH

Take a ceramic object, say a Grecian urn, and throw it hard against a stone wall, so that it shatters randomly into innumerable pieces. Surely such an act can do nothing but create disorder! And yet (so *The Economist* tells us) a strange hidden order emerges. Carefully gather up the pieces of that smashed urn and count the number of pieces larger than 0.1 grams, the number larger than 0.01 grams, and so on, and you will find something remarkable: the pieces will obey a power law. And not just any power law: if it really was a Grecian urn, the exponent will take on a particular value; if it was a ceramic sphere, it will take on another value; and so on. That is, a seemingly disorderly and complex process of fragmentation (which is random growth with a minus sign) produces the simple order of a power law, and the exponent of that power law contains important information – in this case, incredibly, it turns out to reveal the shape of the original object.

Like the idea of order from instability, the idea that simple frequency distributions can be the result of random growth – and that the exponent of the power law is telling us something important about the process – is ubiquitous in physical science. Indeed, we may say that physical scientists now rely on random growth to explain regularities from the ground up, from Bak's earthquake model to the size distribution of meteorites and beyond. Current cosmological models now rely on the random growth in an initial period of "inflation" to generate the uneven distribution of mass that allowed galaxies to form, giving new meaning to the idea of a universal principle. And, if it is good enough for the universe, it is good enough for a trivial thing like the economy!

5

Where We Stand

I started this lecture with what must have sounded like some very peculiar questions. I hope that at this point it is clear not only that the questions make a lot of sense but even what the answers might be. When I asked whether a city is like an embryo or like a meteorite, whether a recession was like a hurricane or an earthquake, I was asking whether the relevant principle of self-organization was that of order from instability or order from random growth.

This has been a pretty strange intellectual journey. I hope you have enjoyed it. Yet many of you may feel that I have been awfully loose. I have made a lot of assertions about models that I have not really fully described, let alone worked out. That was deliberate. I wanted to show you the range of issues, the commonality, and to be frank, the sheer fun involved in thinking about the self-organizing economy. Yet I value disciplined thinking as much as anyone; so in the next part, let us settle down a bit and look at how some real models of self-organization work.

PART
—2—

Self-Organization in Time and Space

In the previous part I tried to suggest, with sweepingly loose exposition, how and why an economy might exhibit principles of self-organization similar to those exhibited in embryos and neural networks, ecologies and hurricanes. Along the way, however, I made some promises that I never quite fulfilled.

First, I said that I would talk about self-organization both in space and in time. In the first part, however, all my examples were spatial. That was deliberate: on just about all counts I am on much more solid ground when I talk about spatial self-organization than when I try to tell temporal stories. Partly that's because the business cycle, which looks as if it might be a self-organizing phenomenon, is also a subject on which it is all but impossible to speak without offending either common sense or the equilibrium macro faithful. It is also true that, if there are striking empirical regularities in business cycles, I am not aware of them. Still, in this part, I will spend some time on the subject, describing two models and one crazy speculation.

Second, I alluded to spatial models without quite explaining what the models were, let alone showing how they worked. So in this lecture I will also take time to explain the workings of three spatial models that illustrate the basic principles of self-organization.

Without further ado, then, let us turn to some of the temporal aspects of self-organization.

6

Dynamics in Self-Organizing Systems

In the field of international trade, there is what might seem to the uninitiated a peculiar distinction between two areas of inquiry: the solid, well-established theory of "growth and trade" and the much more controversial subject of "trade and growth."

What is the difference? "Growth and trade" considers the impact of growth on a trading economy. Suppose that Chinese productivity rises or Japan trains more engineers: what is the effect on world trade patterns and U.S. real wages? These are well-understood issues; what makes them fairly cut and dried is that it is assumed, tacitly or explicitly, that we may ignore any feedback from trade to growth, that the forces of change are exogenous. On the other hand, "trade and growth" is concerned with precisely such feedbacks: how, for example, does trade liberalization affect the rate of technological change? Is a liberal trade regime good for development, or is sophisticated mercantilism better? These are sexy and topical questions, but the answers are far more controversial.

I would argue that there is a similar distinction between two concepts of dynamics in self-organizing systems. One subject, which economists already understand fairly well, involves the time dimension of a system that is self-organizing in some other dimension, such as space. The other subject, which is much more controversial, involves self-organization along the time dimension itself.

Let us start with the easy issue. Imagine an economy that, at any point in time, may be described as a self-organizing system. In particular, imagine that its dynamics form a complex landscape, in the sense I described in the preceding part. And now suppose that this landscape is subjected to exogenous forces of change, forces that gradually exalt the valleys and bring the hills and mountains low. Can we say anything interesting about the likely process of change in the economy?

Yes, we can: in general, a complex dynamic economy will exhibit the pattern that in evolutionary theory is known as *punctuated equilibrium*: long periods of relative quiescence, divided by short periods of rapid change.

Let me illustrate this point, not with the spatial models that I use in most of these lectures but with a model of technology choice. This kills two birds with one stone: I can tell a simple story and at the same time I can demonstrate that the concepts I have been advancing apply to technology as well as location.

Imagine, then, a situation in which households have a choice between two technologies – say, laser discs versus videotape cassettes. Each technology has some inherent advantages over the other, but for many people the most important factor is the availability of some support service, such as movies to rent in video stores. The provision of that service, in turn, depends on the size of the market. So most people will prefer videotape recorders if there are mostly cassettes in the stores, and stores will mostly stock cassettes if that is what most people are prepared to rent.

Figure 6.1 shows what the dynamics of this system might look like. On the figure's horizontal axis is the percentage of stores stocking laser discs (as opposed to cassettes); on the vertical axis the percentage of households with laser (as opposed to videotape) players. The curve H represents the equilibrium share of consumers with laser players, *given* the number of stores offering the discs; the curve R represents the equilibrium share of laser discs available for rental, given the number of households in a position to use them. I assume that each of these shares adjusts gradually toward its equilibrium value.

Finally, I have drawn both H and R as S shaped; that is, I assume that there is a kind of critical mass of available rentals needed to persuade most consumers to buy laser players and a critical mass of consumers needed to persuade stores to stock the discs.

The arrows on the diagram represent the laws of motion. Clearly, there are two stable equilibria in this picture. At one equilibrium, C, there are very few laser players or discs around; in this equilibrium (as in the real world at the time of writing), cassettes are the dominant medium. At the other equilibrium, L, discs are domi-

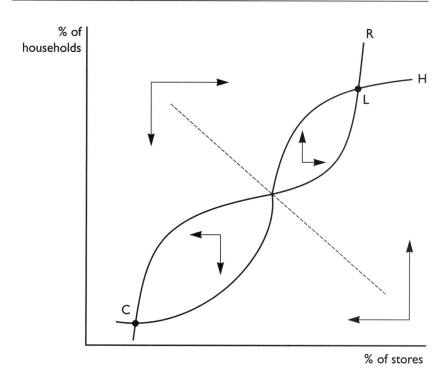

Figure 6.1 Choosing a Video Technology. If consumers choose VCRs
that can play what stores rent, and stores rent what consumers can
play, the choice of a video technology can be self-reinforcing.

nant. So we have the multiple basins of attraction that define a complex dynamic landscape. The broken line illustrates schematically
the division between the *C* basin and the *L* basin.

Now let us introduce change into the picture. Suppose that
although cassettes currently dominate the scene, the direction of
technological change is gradually to increase the advantages of laser
discs. Then other things equal, households will become gradually
more willing to buy laser players; as shown in Figure 6.2, the curve
H will shift gradually up toward *H'*.

But will the actual process of change be gradual? At first it will:
the equilibrium will move along the *R* curve from *C* to *C'*. But at
that point something more drastic will happen: the whole cassette-
dominated equilibrium will collapse. Households will turn to laser
players; as they do so, stores will shift shelf space from cassettes to
discs, further reinforcing the attractiveness of lasers, and so on, leading to a discrete shift of the whole system from *C'* to *L'*.

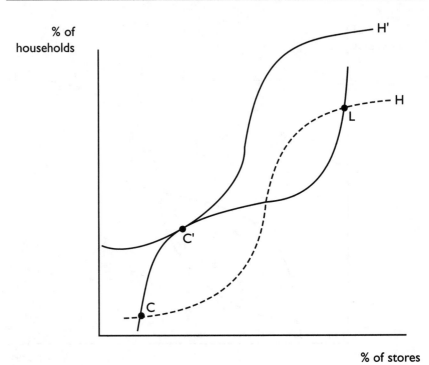

Figure 6.2 Punctuated Equilibrium. Improvements in the less-used technology may have little effect on consumer choices until the system nears the critical point at which the current equilibrium collapses; then a small further improvement will trigger large-scale changes.

The process of change in this story, then, will be one in which gradual shifts in technological advantage usually lead to equally gradual changes in behavior, but in which there are occasional sudden shifts.

I have no idea whether this is a realistic story about the past and future of video entertainment, although it sounds plausible. The point, however, is that the story is quite generic. The axes in Figure 6.1 could represent the numbers of workers and firms in two dif-

---

1. The change in the phase space need not be entirely exogenous; both technological development and evolution are at some level self-generating processes. What is necessary to create punctuated equilibrium is for some processes to proceed much faster than others; for example, basic electronic technology moving gradually, while market shares in videos can be changing much more quickly.

ferent locations (as they indeed did in my 1991 book *Geography and Trade*); or for that matter they could represent the survival strategies of two interacting species, be they competitors, symbiotes, or predator and prey.[1] In any of these cases, we would expect gradual changes in the underlying parameters, and therefore of the phase space, to produce only gradual change in the outcome most of the time, but big changes every once in a while.

The basic logic of punctuated equilibrium should be clear. In a complex landscape, there are many locally stable equilibria; sudden change comes when a previously stable equilibrium becomes unstable, setting the system adrift while it searches for a new equilibrium. But that change from stability to instability always occurs at a discrete moment, no matter how glacial the pace of change in the underlying landscape; and so gradual change in the environment produces dynamics that resemble the traditional description of war: long periods of boredom punctuated by brief interludes of terror.

Does such punctuated equilibrium actually manifest itself in practice? Certainly some technologies, like cellular phones or VCRs, seem to lurk in the background for years then suddenly explode into mass use. Even more notably, urban history contains many examples of explosive growth: Chicago in the mid-19th century, Los Angeles after World War II. And edge cities, when they emerge, often do so with stunning speed: Joel Garreau's book contains "before and after" pictures of Tyson's Corner in Virginia that dramatically make this point.

So the idea that self-organizing systems have a characteristic rhythm of change seems to apply to the economy as well. But in this story the basic sources of change are left unexplained – just as the sources of growth are left unexplained in the "growth and trade" literature. In an ugly, but rather useful piece of jargon, we can say that the dynamics are extrinsic rather than intrinsic.

What would constitute a case of intrinsic, self-organizing dynamics? For that we need a model in which the economy organizes itself in time rather than (or as well as) in space.

---7---

Temporal Self-Organization

TWO QUESTIONS ABOUT THE BUSINESS CYCLE

If you have ever tried to talk about the nature of recessions with an intelligent friend who does not know economics, you have probably discovered that the troubles begin with the way you pose the question. I usually begin by saying something like, "Suppose that there is a fall in investor confidence – let us try to see how that can translate into a fall in output and employment." At this point the friend almost always asks, "But what caused the decline in investor confidence?" and is not easily persuaded simply to take that decline as a given. Indeed, most people seem to think that you are cheating if you start with a decline in confidence, or a monetary contraction, or whatever – *of course*, they think, that leads to a recession, the question is why it happened.

Well, anyone who has followed debates in macroeconomics knows that there is no "of course" about it. It is by no means clear to many economists that a decline in investor confidence will lead to a fall in aggregate demand; if you are a firm monetarist (or is the species extinct?) you believe that nominal income is determined by the monetary aggregates, independent of the state of mind of the business community. In a way, the whole point of Keynes's *General Theory* was to break free of the intellectual straightjacket of the quantity theory, which asserted that changes in investment demand were irrelevant; that is why the theory of liquidity preference, which said that the demand for money depends on the rate of return on alternative assets, played so crucial a role in his thinking. Conversely, however, much of the point of monetarism was to put that straightjacket right back on.

Moreover, it is if anything even less clear to many economists that changes in aggregate demand can have any real effects; even if investor confidence does fall, many academics believe as a matter of principle that any effects of that decline are on the price level rather than output.

Now I am not asking you to take sides in these arguments, though I guess it is pretty obvious that I am a more or less unrepentant Keynesian. The point I want to make is instead one about our profession. For the last 30 years or so macroeconomic debate has focused very heavily on what is sometimes called the *transmission mechanism*, on whether and how a decline in particular types of demands like business investment is reflected in a decline in overall demand, on whether and how a decline in aggregate demand is reflected in a decline in output and employment. Do shifts in the IS curve reduce demand, or is that channel blocked by a near-vertical LM curve? Do shifts in aggregate demand affect output, or is that channel blocked by a near-vertical aggregate supply curve? These are the kinds of questions that have preoccupied macro theorists. And rightly so, because these are deep issues, and we cannot claim to have a real theory of business cycles until we resolve them.

And yet something odd has happened as a result of this focus on the transmission: macroeconomics has increasingly ignored the engine. I recently looked at Greg Mankiw's excellent textbook in macroeconomics, which gives a wonderfully clear overview of the state of the field (you know what that means: it means that I share his views). There is very clear discussion of the reasons why fiscal and monetary policy might or might not have real effects. What is notably lacking, however, is a discussion of the reasons why aggregate demand fluctuates in the first place. Business cycles, in Mankiw's book and in macroeconomics generally, are treated as if they arose entirely from exogenous shocks.

This was not, however, always the way that macroeconomists approached their subject. There was a period that we might call the Keynesian spring, from the war until about 1960, when most macroeconomists accepted Keynes's own view that fluctuations in investment demand were the cause of booms and recessions. They were thus in the position of my lay friend: it seemed obvious to them that a fall in investment demand would cause a recession; the interesting question was what caused the fall in demand. Furthermore, macroeconomists of that era still had fresh in their mind the experience of the Great Depression, in which a massive slump in world demand occurred in the absence of any obvious cause.

So there was a period when economists tried to answer the question that later fell by the wayside: what causes demand to fluctuate? Or to put it my way, how and why does the economy organize itself over time?

NONLINEAR BUSINESS CYCLE THEORY

For a period of a decade or so, a number of economists propounded an explanation of economic fluctuations, generally known as *nonlinear business cycle theory*, that was both intuitively persuasive and analytically elegant. Versions of this theory were propounded by John Hicks and Richard Goodwin in the United Kingdom; there was also an early contribution along similar lines by James Tobin in the United States. Unfortunately, the theory rested firmly on a strongly Keynesian base. When this fell out of fashion, when the monetarist controversy was followed by the debate over aggregate supply, the whole enterprise of trying to explain why and when fluctuations occur rather than how they are possible at all was largely abandoned, and the innovative dynamic theory that had flourished in the 1940s was largely forgotten.

Well, I want to resurrect that theory at least briefly. Even if you think that Keynesian ideas are nonsense, bear with me for a little while, because the point I want to make is less about the nature of the business cycle than about the nature of self-organization.

What, then, was the essence of nonlinear business cycle theory? In its cleanest version (Goodwin's), it envisaged a modified version of the Keynesian cross. As I hope most of you know, the traditional Keynesian cross asserts that planned spending – investment plus consumption – is an increasing function of the level of output in the economy, like the schedule shown in Figure 7.1. But in a closed economy, output equals spending. So equilibrium, defined as a situation where planned spending equals actual spending, must be at a point where the spending function crosses the 45° line.

In Figure 7.1 I have made the conventional assumption that the marginal propensity to spend is less than 1, so that the equilibrium is unique and it will also be stable under any plausible dynamics.

To get from here to a theory of the business cycle, we alter two elements. First, we assume that over some range the marginal propensity to spend is *more* than 1. This can be justified by invoking

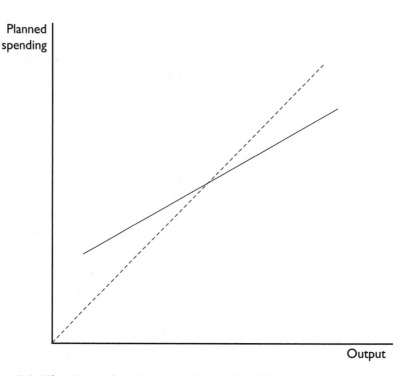

Figure 7.1 The Keynesian Cross. In the standard Keynesian story, the circular relationship between output and spending leads to a unique equilibrium.

investment demand, which could well respond to changes in sales more than one-to-one. However, this strong response of demand to output flattens out at both low and high outputs. If you like, you can suppose that at low output many sectors are doing zero investment and can cut no further, whereas at high output either investment or output itself is limited by capacity constraints. The important thing is that we assume an S-shaped expenditure schedule like the one in Figure 7.2. In this case, of course, there are three equilibria, with the middle one unstable.

Second, we suppose that the expenditure function depends not only on the level of output, but also – negatively – on the capital stock. And we suppose that at the high-investment equilibrium H in Figure 7.2, the economy is accumulating capital, whereas at the low-investment equilibrium L the economy's capital stock is depreciating faster than it is being replaced.[1]

What will happen? When the economy is in a boom, capital will accumulate. This will gradually reduce investment demand at any

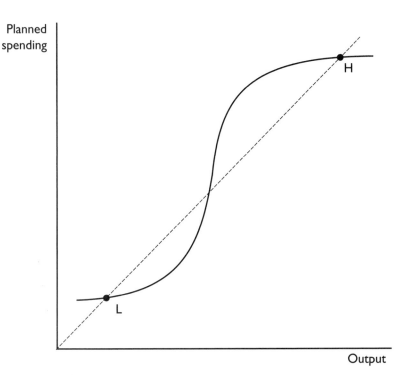

Figure 7.2 Booms and Slumps. If higher output sometimes leads to a
more than one-for-one increase in demand, there may be two
equilibria: one with a high level of output, the other with low

given level of output, shifting down the expenditure schedule. This
process is illustrated by the shift from S_3 to S_2 to S_1 in Figure 7.3.
And the consequence is obvious: at a certain point the boom will
collapse. Once the economy is in a slump, the capital stock will
begin to decline and investment will begin slowly to revive; the
schedule gradually will shift back up, from S_1 to S_2 to S_3, and the
boom will be on again.

The rise and fall of the capital stock plays a key role in the story,
but that role is hidden in Figure 7.3; it can be revealed in a phase
diagram like Figure 7.4, in which the capital stock is on the horizontal axis, the level of output on the vertical. The story about the
business cycle that I just told implies that the economy ends up fol-

---

1. Or more plausibly, we add an exogenous component of expenditure that
grows with the economy's potential output. What declines over time is then the
ratio of the capital stock to this trend.

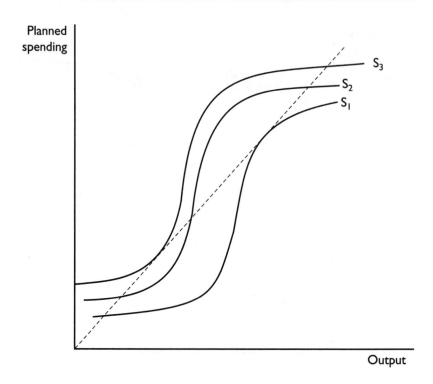

Figure 7.3 The Business Cycle. The accumulation of capital during a
boom eventually triggers a slump, while the depreciation of capital during a slump sets the stage for the next boom.

lowing a limit cycle like the one shown in the figure, alternating
periods of high output and rising capital with periods of low output
and falling capital, each phase of the cycle setting the stage for the
other.

The limitations of this story are pretty obvious, but I have
described it to make two points.

First, I describe nonlinear business cycle theory partly to uphold
the honor of my profession. If you read books like Mitchell
Waldrop's *Complexity: The New Science at the Edge of Order and Chaos*,
you get the impression that economists are dull people who never
thought of trying to introduce nonlinear dynamics and self-organization into their analysis until the visionaries at Santa Fe pushed
them. Well, nonlinear business cycle theory is more than 45 years
old. If it failed to catch on at the time, it was not because economists
were too unimaginative to realize the possibilities, it was because

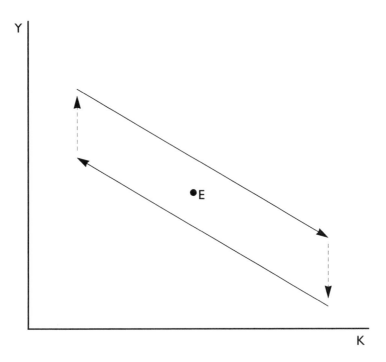

Figure 7.4 Another View of the Business Cycle. The same cycle can be seen in phase space. There is always an equilibrium without a cycle, but it is unstable.

they were, understandably, dubious about some of the approach's economic foundations.

Second, and more important, I want to point out the parallels between the reasons for temporal self-organization in nonlinear business cycle theory, and those in my edge city story.

The limit cycle in Figure 7.4 is an equilibrium of the dynamic system. But, is it the only equilibrium? No, there will always be a stationary equilibrium, in which the levels of output and the capital stock are just right to support a level of investment that is just enough to offset depreciation (and trend growth, if you want to make this a growth-and-cycles story). I have indicated this constant-output, or more likely steady-growth, position as E in the figure. Given that there exists an equilibrium in which output does not fluctuate, then, why do we focus on the fluctuating equilibrium? Because the stationary equilibrium is *unstable*. Jiggle the economy, and it will soar into a boom or plunge into a slump, and the business cycle will commence.

Nonlinear business cycle theory, then, can be thought of as an approach in which temporal order arises from instability; business cycles, in this view, are like convection cells, hurricanes, or for that matter edge cities.

We can push that last parallel a bit further. I argued in the first part that the existence of multiple urban subcenters depends on a tension between centripetal and centrifugal forces, with the forces of attraction fading more rapidly with distance than the forces of repulsion. If there were only centripetal forces, everything would tend to form one big clump. There is a similar logic to the repetitive nature of the nonlinear business cycle. It is possible to imagine a model in which firms want to invest if others do, and that is the end of the story; in such a model there would be high and low level equilibria, but no cycle. (The Big Push model of Murphy, Shleifer, and Vishny, 1989, is a good example.) What drives Goodwin's cycle is the way that investment encourages investment in the short run but discourages it in the long run. In his model, firms tend to raise or cut their investment at the same time, because the level of investment determines the level of aggregate demand, which drives the level of investment. This is the temporal equivalent of localized agglomeration effects. A prolonged period of high investment, by increasing the capital stock, eventually discourages further investment; this is the temporal equivalent of centrifugal forces like land rent and congestion.

The largely forgotten theory of nonlinear business cycles, then, is quite close in underlying structure to my story about edge cities; both are stories of self-organization based on the principle of order from instability.

Of course, nonlinear business cycle did not make it into the economic canon. I have given some reasons for that failure, but let me give another: the approach seems to predict too regular a cycle. The model literally predicts a cycle that would tick like a macroeconomic clock, but we have little difficulty in imagining that gradual changes in structure, plus exogenous shocks, would make the timing of the cycle acceptably irregular. No, the problem is that nonlinear business cycle theory suggests that booms and slumps should be all more or less the same *size* – the economy should always surge from the floor to the ceiling, then plunge down again. In fact, however, booms and slumps seem to come in all sizes. How is this possible?

PERCOLATION ECONOMICS

Recently Jose Scheinkman and Michael Woodford applied the principle of order from random growth to economic fluctuations. Their model was inspired by the work of Per Bak; I am indebted to their work for making it clear to me what Bak and his concept of self-organizing criticality is all about.

The Scheinkman–Woodford model envisages an economy whose input–output structure consists of a large number of "layers." Firms producing final goods form the top layer; they purchase intermediate inputs from the next layer, which in turn purchases inputs from the next layer, and so on. Firms at all levels hold inventories of their own products, and they follow an "S-s" rule on production: when an order for one unit comes in, if they have inventory, they satisfy the order by drawing that inventory down. If they do not have inventory, they produce *two* units – which requires that they place orders for inputs from two upstream firms – and put the extra one into inventory. Orders for final goods arrive randomly.

Obviously an order for a final good will sometimes produce a chain reaction of orders for intermediate goods. But how far will this chain reaction spread? It depends on the levels of inventory. If inventories are high, orders will typically be met out of stock and will not generate additional orders. If inventories are low, each order will typically give rise to two more orders. Loosely speaking, there should be a critical level of inventories below which the expected length of chain reactions becomes infinite.

Now comes the key insight: the level of inventories will always tend to be near that critical level. Suppose that the level of inventories is low. Then an order for a final good will typically set in motion a long chain reaction of orders for intermediate goods; because each firm that is obliged to produce to satisfy its order also adds a unit to its inventory, the level of inventories will tend to rise over time. Suppose, on the other hand, that the general level of inventories in the economy is high. Then most orders will typically be met from inventory – chain reactions will die out quickly and thus the level of inventories will tend to decline. So the economy will tend to evolve to the edge of the critical level of inventories.

At a formal level, what I have just described is a familiar story in physics: it is a problem in *percolation theory*. The classic question in percolation theory, which gave the field its name, was the following: how far will water penetrate into a porous rock? It turns out that a good way to model this is to think of the rock as a lattice of holes, with some but not all neighboring holes connected. Percolation theory typically thinks of each potential connection as being either open or closed with some probability p, then studies the size distribution of connected regions. As one might expect, there is normally a critical probability at which the expected size of connected regions becomes infinite. Less obviously (indeed, somewhat mysteriously), just about all percolation models have the feature that when p is close to its critical value, the upper tail of the size distribution of connected regions follows a power law; that is, the frequency of regions larger than any given size S is proportional to S^{-a}, with a depending on the shape and dimensionality of the lattice. The point seems to be that, if we think of water penetrating into the rock from a particular hole, when p is close to its critical value this spreading becomes a process of random growth whose rate is independent of the volume already reached and thus generates a distribution of "objects," connected regions filled with water, that obeys a power law.

It may not be immediately obvious that Scheinkman and Woodford's model of the economy resembles a porous rock, but in fact it does. Think of firms as the holes in the rock, with the connection between two firms closed if the upstream firm has a unit in inventory, open if it does not. Then the question of how far a chain reaction spreads is equivalent to the question of how far a drop of water penetrates into the rock.

And, if the probability that two holes will be connected is close to the critical value, then the size distribution of connected regions, of the chain reactions that result from arrivals of final demand, will be described by a power law. That is, even an almost steady flow of demand will generate fluctuations in production of all sizes.

The physicist Per Bak has argued that many physical and social phenomena can be modeled as percolation systems that naturally tend to move to the edge of criticality. Bak claims, in particular, that the concept of "self-organized criticality" explains such striking empirical regularities as the Gutenberg–Richter power law relating the sizes of earthquakes to their frequencies.

The attractiveness of an approach like this for business cycle theory is obvious: it explains how the economy can endogenously generate fluctuations without having the kind of typical size of fluctuation implied by the old nonlinear business cycle theory. The disadvantage is that it seems, if anything, even less realistic, makes less contact with what seems to happen during a boom or slump, than the old theory. Nor is there, to my knowledge at least, any evidence that the size distribution of business fluctuations actually obeys a power law.

What *does* obey a power law remarkably well, as I already pointed out in the first part, is the size distribution of metropolitan areas; and we will come back to that example shortly.

Meanwhile, however, back to the business cycle. I am agnostic about both the Hicks–Goodwin and Scheinkman–Woodford fluctuation theories. I admire them both and find both a little hard to swallow as empirical propositions. At this point I prefer to regard them both more as illustrations of how one might approach self-organization in time than as finished statements of how one actually ought to do it.

And yet I cannot resist the temptation to try to use these approaches to explain something strange about the real world.

PHASE LOCKING AND THE GLOBAL BUSINESS CYCLE

Global interdependence is a favorite phrase these days for politicians who want to sound sophisticated. Indeed, the *Wall Street Journal* once published a whole article on the anthropology of Multilateral Man, that special tribe that seems to like nothing better than to attend G7 meetings, Bank-Fund meetings, economic summits, and so on.[2] And yet it is a truism among those who actually work with quantitative models of international macroeconomics that global interdependence, at least as far as macroeconomic linkages are concerned, is quite modest.

The arithmetic is straightforward. The United States and the European Union each export about 2 percent of their output to each other. There is reasonably solid econometric evidence suggesting that the income elasticity of import demand is around 2. So consider the often stated view that economic recovery in the

---

2. According to the *Wall Street Journal*, the two basic tenets of Multilateral Man are that cooperation must be improved to facilitate coordination and coordination improved to facilitate cooperation. (Actually, that sounds a bit like a self-organizing system!)

United States helps Europe, by stimulating European exports. Well, according to the arithmetic, a 1 percent rise in the U.S. GDP will increase Europe's exports to the United States by 2 percent. That is a stimulus, all right – but it is only 0.04 percent of European output, pretty small change considering the rhetoric.

And yet there is a paradox. If interdependence is so limited, why do we so often have global recessions that strike many of the world's nations at the same time? Why did all of the industrial world share in not only the Great Depression but in the major recessions of 1974–1975, 1979–1982, and 1990–1992?

One of the luxuries of a format like this one is that I can include the kind of loose speculations that I could never write in a journal and that I can explain, as I am doing now, that I do not necessarily believe in the theory I am advancing. So here is a crazy idea about the global business cycle: it is an example of "phase locking."

Phase locking, otherwise known as *mode locking, frequency pulling,* or simple *synchronization of two oscillators*, is one of those phenomena that occur in wildly different contexts and at very different scales. It was apparently first described by Huygens in the 17th century, who reported that two pendulum clocks placed back to back on the wall separating two rooms would synchronize their motions. The same principle turned out to have a more cosmic significance when it was learned that some asteroids have orbital periods that are related to that of Jupiter by the ratio of two integers; for example, Pallas has an orbital period 7/18 that of the giant planet.

It seems to me that phase locking might well occur in an international version of either the Hicks–Goodwin or the Scheinkman–Woodford business cycle. Start with the Hicks–Goodwin cycle. Suppose that you envisage two economies with modest linkages, each of which in isolation would have a cycle like the one illustrated in Figure 7.4, with the length of the cycle not too different. Now suppose that one economy plunges into slump at a time when the other economy has accumulated almost enough capital to slump as well. One can easily imagine that the large slump abroad could provide the small adverse shock needed to end the boom at home. Similarly, a country that is not too far from a spontaneous recovery might well be triggered into a somewhat premature boom by a large surge in foreign markets. Like the two back-to-back clocks that started ticking in unison, the two economies would not need to be very strongly linked to develop a synchronized cycle; a modest linkage would do as long as they were predisposed to have cycles in any case and had fairly similar natural periods.

What about the Scheinkman–Woodford cycle? Imagine two economies in which, as in their model, firms purchase inputs from successive layers of upstream suppliers; but imagine that there is some linkage in the sense that with some modest probability one of the upstream suppliers for any given firm will be in the other country. Following the logic of their story, each of these two economies will tend to evolve to the edge of criticality, so that the arrival of an order for even a single unit of the final good will sometimes trigger a large cascade of induced orders for intermediate goods.

Suppose, however, that there is a very large cascade. This will generate a fairly large number of orders in the other economy, even if the linkage is modest, and will therefore have a pretty good chance of starting a large cascade in the other country as well. It seems to me that even with fairly small "import shares," one might well find that large economic fluctuations tend to occur in both economies at once.

Anyway, all of this is a somewhat wild speculation. I include it because I think it is interesting and also to illustrate the way that thinking about self-organization stimulates you to think about familiar issues in novel, if not always sensible, ways.

The subject of self-organization of economies in time is a fascinating one, and I am sure that I have barely touched its surface. Nonetheless, that is all I am going to say on the subject. Instead, at this point I want to return to space and explain at greater length why and how the models I alluded to in the first part work.

—————— 8 ——————
Models of Spatial Self-Organization

The writer Robert Benchley once explained that there are two kinds of people in the world: people who divide people into two kinds, and people who do not. I am one of the people who do; that is, I believe very strongly in the use of simplified models to help us think about social science.

In this book, I am following the usual, dishonest practice of scientific exposition: in the first part I presented some intuitive concepts in a breezy, isn't-it-obvious tone; only now am I about to elaborate on those concepts with a set of more fully worked-out models. There are good reasons for presenting ideas this way – it is a much more pleasant experience for the listener or reader than a direct plunge into the models, and therefore it is a much better way for someone with an idea to hold the audience's attention. Why, then, do I call it dishonest? Because it is completely misleading in the picture it gives of how ideas are formulated or how one tells good ideas from bad.

The truth is that interesting ideas almost never start with clear but informal intuition, which is then backed up by a formal model; intuition that comes that easily usually is not very interesting. The actual process is far more confused. Here is how I do theory: I start with a vague, murky intuition, much of which turns out to be wrong.[1] I then try to get a grip on that intuition by building a formal model – a model that, on first pass, is usually an awkward thing

---

1. Here is an example of what I now believe to be a wrong intuition. It used to seem to me that, to explain a tendency to form multiple agglomerations at some distance from each other, one must suppose some mechanism whereby a successful location casts an "agglomeration shadow" over its neighbors, much as a large tree prevents smaller nearby trees from getting sunlight. Well, this turns out to be

full of pointless complications. Then I begin a sort of process of intellectual self-organization, in which the model suggests a revision of my intuition, the revised intuition suggests a way to simplify the model, and so on. In the end, if I'm lucky, I end up with a story that seems obvious to other people as soon as they hear it. But it was not at all obvious to begin with, and the apparent obviousness is the result of a lot of hard grinding away on the formal model.

All of this is by way of explaining that, after the whirlwind tour of self-organization theory, it is time to settle down and do some homework. In the remainder of this part I'm going to try to explain the workings of three formal models of spatial self-organization. The first is the "edge city" model described in the first part. The second is a "self-organized" version of central place theory; that is, an attempt to show how a Lösch-type lattice of central places could arise spontaneously through an invisible hand process. The third is Simon's model of the power law on city sizes. Of these, the central place model is by far the most difficult and least realistic. It was also the model with which I started – as I said, the course of theorizing is not as smooth as later rewriting can make it seem. And as you shall see, there are reasons for including this model even though it is neither realistic nor easy to understand.

By the way, those of you who find it hard to sit through equations should not leave. I am going to try to explain how these models work with only a few equations and figures; the heavy stuff is reserved for the appendix. The truth is that I suffered through all of this in reverse order – mathematical grinding first, loose mathematical intuition next, grand vision last. But out of the goodness of my heart I will not make you share my pain.

So let us begin with a look at the underlying logic of the edge city model.

THE EDGE CITY MODEL

The purpose of what I am calling the edge city model is to explain, in as minimalist a way as possible, how the interdependent location decisions of businesses within a metropolitan area could lead to a

---

all wrong: as we've already seen (Figure 1.5), widely spaced agglomerations can emerge from a process in which neighbors initially seem to grow together. The actual reason why waves turn into spikes is quite different and explained below.

polycentric structure, in which business is concentrated in several spatially separated clusters. It is my desire to explain the emergence of a polycentric pattern, rather than just a single agglomeration, that gives rise to some subtleties – otherwise it would be just a matter of agglomeration economies giving rise to agglomeration. Here, however, I am trying to explain how a tension between centripetal and centrifugal forces creates a spatial pattern.

To simplify matters, we consider a one-dimensional city, with households and businesses strung out along a line. We will want to move back and forth between imagining that the line is actually a circle, so that there are no ends, or that it is simply very long, so that the ends are very far away from most places. Realistically, we should also think either of the line as consisting of some number of discrete locations or of businesses dotted along the continuous space. In the spirit of physics, however, I will do neither and instead think of representing the distribution of business along the line as a continuous *density*, with $\lambda(x)$ the density at position x.

At any given time, the desirability of locations will differ. Because this is supposed to be a model of self-organization, I assume that there are no inherent differences among locations. Instead, desirability depends only on the distribution of businesses themselves. I suppose in particular that a concentration of businesses at some location z exerts both positive and negative effects on the desirability of location x, with both the positive and negative effects fading out with the distance D_{xz} between the two locations:

$$P(x) = \int_z [A(D_{xz}) - B(D_{xz})]\lambda(z)dz \qquad (8.1)$$

In the language of economic geographers, $P(x)$ may be described as a "market potential" function. As you might guess, the results of the model depend crucially on the way that those positive and negative effects fall off with distance. The simplest, and it turns out most productive, assumption is that they fall off exponentially with distance, with A falling off faster than B. It will be useful, however, also to imagine a situation in which the positive or negative forces between businesses are constant up to some distance, then suddenly disappear. In this case, ranges of positive and negative spillovers between businesses would be well defined. Of course, this case can be viewed as an approximation to the case of gradually declining effects, which is actually the way that I will treat it.

Because locations differ in desirability, businesses will have an incentive to move. The easiest way to model that moving decision is simply to assume that businesses move gradually away from undesirable locations and toward desirable ones. If you want to write an equation specifying that movement, you have to be a bit careful, to be sure that the rules imply that the total number of businesses moving in to better than average locations equals the number moving out of worse than average places. Here is a particular rule that works. First, define the average desirability – actually, the desirability of the location occupied by the average business – as

$$\overline{P} = \int_x P(x)\lambda(x)dx \tag{8.2}$$

Now assume that the density of firms at each location changes according to the rule

$$\frac{d\lambda(x)}{dt} = \gamma[P(x) - \overline{P}]\lambda(x) \tag{8.3}$$

(What is that extra $\lambda(x)$ doing in there? Check the algebra and you will see that I need it to make sure that the *total* change in the number of firms equals 0.)

And that is it – a minimalist model indeed. But how do we analyze it?

Fluctuations and Instability

One way to analyze this model is simply to put it on the computer and simulate it. Take some number of locations around a circle – 12 or 24 are good numbers, because they are fairly small numbers with a fairly large number of divisors – make up some parameters, use some rule that includes a random element to create an initial spatial distribution of businesses, and see what happens. That is not at all a bad strategy. Indeed, the willingness to engage in computer-assisted thinking, to do theory without theorems, is a hallmark of the emerging literature on self-organization.

When you simulate the model, you will find that for a wide range of parameters our abstract metropolitan area does indeed engage in a process of self-organization. Even if you create an initial spatial distribution of businesses that differs only imperceptibly from perfect flatness, all businesses eventually end up in just a few locations.

Moreover, if you change the parameters you will find that the results change in sensible ways. Make the positive spillovers stronger, and there will be fewer, larger business districts; make the negative spillovers stronger, and the districts get smaller and more numerous. Make spillovers fade out faster with distance, and business districts tend to be closer together; and so on. In a rough way the simulation approach does seem to convey an understanding of the process of self-organization.

And yet after doing a number of simulations one finds oneself driven back to paper and pencil. Partly this is because even those of us who believe that numerical methods are perfectly legitimate still feel that we want something more – that we want to have models of our models. But there is more to it than that. When you look at simulation results, you get a clear sense that some kind of hidden principle of order is lurking in this model about which you ought to be able to offer more than vague intuition and the evidence of computer runs.

The sense of a hidden principle emerges clearly from Figures 1.5 and 1.6. Each of these figures shows the results of a simulation in a model city with 24 locations around a circle, with different parameter values. (In fact, the only difference is that spillovers decline twice as rapidly with distance in Figure 1.6.) In each case, the initial allocation of business to locations was done under a rule that ensured both that the starting distribution was very flat and that deviations from that flat distribution were random. And yet, not only do the runs show self-organization, they show self-organization into perfectly regular structures. In Figure 1.5 the two business districts are at locations 8 and 20; that is, exactly 12 apart, which is to say on opposite sides of the circle. In Figure 1.6 there are four business districts, at regular intervals around the circle. Let me now assure you that you can run the model over and over again with these parameters, with a different random distribution of business each time, and you will get the same results. That is, for the parameters in Figure 1.5 you will always get two business districts exactly opposite one another, for the parameters in Figure 1.6 you will always get four equally spaced districts. (Of course, which locations play those roles changes – the concentrations may be at, say, 1 and 13 instead of 8 and 20.)

Where do these regularities come from? To understand that, we need to use a trick very familiar to physical scientists, engineers, and some econometricians, but that has had very little application in economic theory so far: we need to think of the spatial distribution of business as a Fourier series.

The basic idea of Fourier series – and the basic idea is all that we will need – is that an irregular wiggle can be decomposed into the sum of a number of regular, sinusoidal wiggles at different frequencies. If we think of $\lambda(x)$ as a wiggle along a line of length L, then it can be decomposed into nice, regular sine waves at frequencies 0 (a horizontal line at the average value of λ), $2\pi/L$, $4\pi/L$, $6\pi/L$, and so forth.

Now those regular fluctuations at different frequencies are not really there – they exist only in the mind of the theorist – but nonetheless the regularity of the fluctuations is what explains the surprising order in the simulation results.

How can this be? The key point is that fluctuations at different frequencies will tend to *grow at different rates.* More specifically, very high frequency fluctuations will tend to go nowhere in particular, and very low frequency fluctuations will tend to fade away. Only in some middle range will the flucutations grow; and some preferred frequency will grow fastest. What happens is that this preferred frequency comes to dominate the distribution; the regularity of the fluctuation at that frequency creates order in the self-organizing city.

This sounds faintly mysterious; so let us take it step by step.

The key insight here is that fluctuations at different frequencies have different growth rates and that (subject to certain conditions) the fastest growth rate is at some intermediate frequency. In the appendix I show that this is true when the falloff of positive and negative spillovers takes a smooth exponential form. Here I restrict myself to showing that it is true given the simpler "range of influence" story, in which effects extend without any falloff up to a certain distance, then suddenly drop off to nothing. In fact, let us make life even simpler: let us assume that the range of influence of negative spillovers is precisely three times that of positive effects! You will see why that is convenient in a second; I hope that you will also be convinced that the basic insight remains true more generally.

To ask whether and how fast a fluctuation grows, we imagine a city in which the distribution of business can in fact be described as a regular, sine-wave fluctuation around the average density. How can we do this when even the Fourier series representation gives us a city with many fluctuations at different frequencies? As long as the fluctuations are small, we can approximate our model by a set of linear equations; and as long as we can do this, we can, as it were, partition our metropolis into a set of independently evolving shadow metropolises – one with a wavelength of 40 miles, one with a wavelength of 20, and so on.[2] The linear approximation eventually breaks down, but we'll come to that later.

Let me now try to convince you of three propositions. First, very high frequency fluctuations will grow at best very slowly over time. Second, very low frequency fluctuations will actually fade out over time – as long as the centrifugal forces are strong enough relative to the centripetal forces. Finally, fluctuations at some intermediate frequency will grow fastest.

To think about this, imagine in each case that the fluctuation has a peak at location 0 (we can relabel locations to make this true). Given this, is location 0 a desirable or an undesirable place for businesses to locate? If it is a desirable location compared with the average location, the density of businesses at location 0 will grow over time, which means that the fluctuation at that frequency will grow. If 0 is an undesirable location, the density of businesses there, and thus the size of the fluctuation, will shrink.

The desirability of a location depends, however, on the density of businesses in other locations. Above-average densities of business within the range of positive spillovers will tend to raise market potential of location 0; above-average densities within the larger range of negative spillovers will tend to lower that potential.

So let us look at some pictures of fluctuations at different frequencies. Figure 8.1 shows a high frequency fluctuation, with a number of peaks and troughs within the ranges of both positive and negative spillovers. Given such a fluctuation, will 0 be an especially desirable or an especially undesirable location? If you think about it, you will realize that the answer is neither. Because there are so many peaks and troughs within the ranges of spillover, favorable and unfavorable effects will roughly cancel each other out. This means that

---

2. To be more precise: any sinusoidal fluctuation around the flat equilibrium turns out to be an eigenfunction of this dynamic system; its growth rate is the associated eigenvalue. And the whole process I am describing is precisely one of representing a dynamic spatial model as a partial differential equation, whose solutions are all sine and cosine functions.

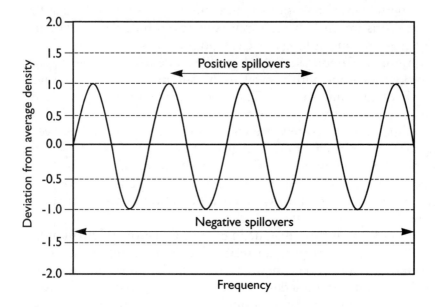

Figure 8.1 A High Frequency Fluctuation. Positive and negative spillovers will nearly cancel out, preventing any growth.

there will be little tendency for additional business to move to location 0; this high frequency fluctuation will not tend to grow much over time.

By contrast, Figure 8.2 shows a very low frequency fluctuation. Here all the relevant business densities – those that occur within both the range of positive spillover and within the range of negative spillover – are well above average. But if the negative spillovers are strong enough – a condition that is easy to derive in any specific model – then the net effect of these spillovers will be to make 0 an *undesirable* location, a location with below-average market potential. So a low-frequency, long-wavelength fluctuation like this will tend to die out over time.

Finally, Figure 8.3 shows an intermediate case: a fluctuation that just happens to be the right frequency so that all of the locations within the range of positive spillovers have above-average business density, all those within the range of negative spillovers that extends beyond this zone have below-average business density. (Remember, I have assumed for the moment that the range of negative spillovers extends precisely three times as far as the range of positive effects; now you see why.) In this case, it is unambiguously clear that 0 is a

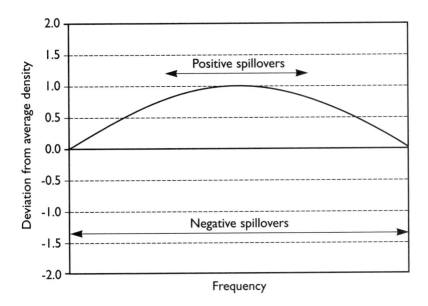

Figure 8.2 A Low Frequency Fluctuation. Negative spillovers will dominate causing the fluctuation to shrink.

location of above-average market potential. Consider that, throughout the range of positive spillovers, the above-average density contributes positively to market potential at 0. Moreover, once we go
beyond to the range of negative spillovers, the business density is
below average – and this *also* contributes to the desirability of location 0.[3]

So a fluctuation at this frequency will grow over time. This picture actually works only for the special case in which the market
potential function is characterized by limited ranges of effects and in
which the range of negative effects is precisely three times the range
of positive effects, but the intuition holds up more generally.
Provided that centripetal and centrifugal forces are not too uneven-
ly balanced, low frequency fluctuations will die out, high frequency fluctuations will go nowhere in particular, but fluctuations at an
intermediate frequency will grow and some particular frequency
will have the fastest rate of growth.

---

3. Strictly speaking, there is a negative effect from the concentration of business in the middle, which is necessarily offset, two to one, by the effects of the
troughs on either side.

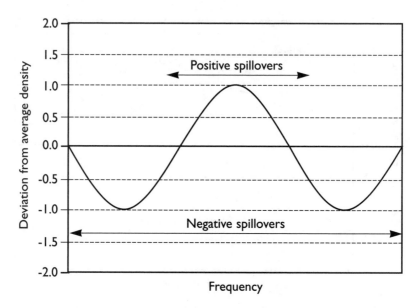

Figure 8.3 An Intermediate Frequency Fluctuation. If a fluctuation has the right frequency, it will grow over time, eventually dominating the economy's spatial pattern.

We can now start to see how a regular pattern may emerge from a disordered initial condition. Suppose that the initial distribution of businesses is almost, but not quite, uniform. We can represent the deviation of that distribution from flatness as the sum of a series of regular, sinusoidal wiggles; but if the distribution is almost flat, these wiggles will be very small. Over time, some of the wiggles will grow; if this period of growth lasts long enough, the distribution will become increasingly dominated by whichever wiggle grows fastest, no matter how small it may have been to start with.

Indeed, it is possible to show numerically that this is precisely what is going on in the examples illustrated by Figures 1.5 and 1.6. In both examples, I set up a circular city of circumference 2π. The simulations that produced the figures were based on the full, non-linear model. It is easy, however, to take a linear approximation to the model in the vicinity of perfect flatness, and to calculate the implied growth rates of fluctuations at different frequencies. Table 8.1 shows the results of that calculation. For the parameters used to generate Figure 1.5, a fluctuation at frequency 2 grows fastest in the linearized model; for the parameters used to generate Figure 1.6, a fluctuation at frequency 4 grows fastest. And sure enough, the first

Table 8.1

| | Growth rate: | |
Frequency	Two-center case	Four-center case
1	–.1991	–.3796
2	.0223	–.0215
3	–.0081	–.0228
4	.0096	.0058
5	–.0017	–.0034
6	.0047	.0049
7	–.0007	–.0006
8	.0027	.0033
9	–.0004	–.0003
10	.0018	.0023

Predicted growth rates in small city model.

set of parameters consistently yields two edge cities facing each other, the second consistently yields four equidistant concentrations.

But wait a second. Some of you may be saying, "You've been telling us a story about nice, smooth, sinusoidal fluctuations. What happens in Figures 1.5 and 1.6, however, is that all business concentrates in only a few locations; we end up not with wiggles, but with spikes (or, if you look at the whole space–time picture, we get dorsal fins rather than rolling ridges). How can this actually sharp-edged picture emerge from a story about gradual divergence?"

From Wiggles to Spikes

To understand the spikiness of the eventual distribution of business, we need to think a bit more carefully about an issue that I have slipped past in my exposition so far: there is an adding-up constraint in this model. I am assuming that the total number of businesses, or equivalently the average density of business, remains unchanged. This constraint can be ignored during the early phase of self-organization, in which fluctuations away from average density remain small, but it starts to play a crucial role at a later stage, when they have become larger.

Why can we, in effect, ignore the adding-up constraint during the early stage of self-organization? Because a growing fluctuation, being a sine wave around the average business density, by definition has as much area below the line as above it. Therefore, its growth has no effect on the average density. If we think of the distribution of business as the sum of fluctuations, and we then imagine some of these fluctuations growing while others shrink, the adding-up constraint will be satisfied by each fluctuation and therefore automatically satisfied by the model city as a whole.

But this process cannot go on forever, for one reason: the business density at any location cannot become negative. A growing sine wave will sooner or later imply negative values greater than the average density; at this point, the process that I have described must break down.[4]

To see why this means that wiggles become spikes, look at Figure 8.4. To understand this figure, imagine that the deviation of the density of business from its average level, here normalized to equal 1, has become completely dominated by a fluctuation at some one frequency, and this fluctuation keeps on growing. I show what happens when the sine waves can grow no further without implying negative values. Clearly, the sine wave starts to develop "flat bottoms" – ranges where the density of business is zero. But these flat bottoms in effect truncate part of the below-average part of the sine wave; if the curve retained its shape, there would be more distribution above average than below, which is possible only for the children of Lake Wobegon. Obviously, then, something must give: the growing peaks must grow, not only at the expense of the troughs, but by pulling in businesses even from locations that had been growing up to that point. As a result, the peaks get sharper. And this process continues until the wiggles become spikes.

I think – I hope – that this image is clear. It may even seem obvious. But there is a fairly subtle feature about the later stages of self-organization in the edge city model that may be worth emphasizing.

Let us ask the following question: Why does each concentration of business in this model end up surrounded by a clear area without businesses? Why, in the real world, do our metropolitan areas develop reasonably compact business subcenters, rather than the relative-

---

4. Actually, given the dynamic rule I have assumed, there is no sudden breakdown; rather, the nonlinearity built into the dynamics becomes more and more important as the distribution of business diverges from flatness.

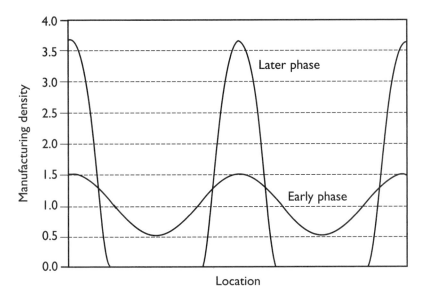

Figure 8.4. From Wiggles to Spikes. As the spatial distribution of activity gets increasingly uneven, it starts to change from a smooth sine wave to a pattern of widely separated spikes.

ly even sprawl that many people envisaged a couple of decades ago? It seems natural to think that a center surrounded by a void must be directly interfering with the development of nearby areas – that, to use the felicitous phrase proposed by Brian Arthur, it must exert an "agglomeration shadow." In fact, that is not at all what happens here. The areas close to the emergent concentrations have higher market potential than places farther away. They just do not have high enough market potential to compete for the limited pool of businesses. The spikiness of the eventual distribution of business is a consequence of the adding-up constraint. That is, it is a general equilibrium effect.

This, then, is how what I have called the *edge city model* works. I hope you share at least some of my satisfaction at seeing how a very simple model can yield complex and subtle dynamics – and conversely, how dynamics that at first sight appear hopelessly complicated turn out to be governed by a surprising principle of order. Moreover, I believe that the model is basically right as an account of how metropolitan areas engage in spatial self-organization.

Yet there is one big objection to this model: although the conclusions do not follow in a trivial way from the assumptions – it is not simply a matter of agglomeration economies producing agglomerations – the model is still very much in a reduced form. I assume that there are centripetal and centrifugal forces that decline with distance, but what are these forces?

It is possible to speak loosely about what they must be. Joel Garreau, whose book *Edge City* has been a useful guide throughout this book, attributes the rapid growth of subcenters once they reach the "critical mass" of 5 million square feet of office space essentially to the growth of local business services; for example, the ability to support a local luxury hotel. He also stresses the role of land costs and traffic congestion in limiting the growth of edge cities. So these forces are not mysterious. But so far I have not explicitly modeled them.

Let us talk about how we might fill that gap.

A CENTRAL PLACE MODEL

A few years ago I gave a set of lectures on the subject of economic geography. The little book that came out of those lectures, *Geography and Trade*, had at its core an approach to location that in effect derived centripetal and centrifugal forces from microfoundations; that is, spillovers themselves became an emergent property.

Here is the basic approach. I envisaged an economy in which there was an immobile, geographically dispersed sector, "agriculture," and a mobile sector, "manufacturing." Manufacturing consisted of many differentiated products, with economies of scale at the plant level. (The market structure was therefore monopolistically competitive.) There were transport costs for manufactured goods, though for convenience I assumed away any transport costs for agricultural production. There were no pure externalities, no direct spillovers between firms or households.

How can such a model generate agglomerations? The answer lies in the interactions among economies of scale, transportation costs, and factor mobility. Because of economies of scale, each product will be manufactured in only a few locations.[5] The favored locations will tend to be those with good access to markets and with good access to the goods produced by other plants – in the language of

---

5. Strictly speaking, in the monopolistic competition model each individual variety is produced in only one location.

development economies, preferred sites will be those with good backward and forward linkages. But where will a firm find good access to markets? Precisely where many other firms also choose to locate. Thus there is a circular process whereby firms tend to concentrate near other firms, producing agglomerations. The agglomeration economies are not assumed; they are derived. Furthermore, they are a purely emergent property. There is no immanent centripetal force at the level of the individual firm; things that we can call *forward* and *backward linkages* appear when large numbers of firms interact.

Will a model along these lines always produce a single agglomeration? No, against the centripetal forces must be set a centrifugal force. Firms provide each other with markets, but they also compete for markets, in particular the market provided by the dispersed agricultural population. The tension between centripetal and centrifugal forces at least suggests the possibility of an emergent pattern with multiple centers.

You can guess the next step. Let us simulate it on the computer. Begin with an almost but not completely smooth distribution of manufacturing on a — you guessed it! — circle, and simply let the model evolve according to some simple dynamic rule. It turns out that the simulation is much more complex in this case than in the edge city model, because you must solve a general equilibrium model at each step: given the spatial distribution of manufacturing, you must solve a set of simultaneous nonlinear equations that determine wage rates and prices at all locations, use the result to determine the incentives for firms to move, and so on. Nonetheless, it can be done. And, as long as neither the manufacturing share nor the degree of scale economies is too large, the result is the emergence of a regular pattern of evenly spaced manufacturing centers.

Figure 8.5 shows the results of one such run, on a circle with 12 locations. The X axis shows the locations; the Y axis shows time; the Z axis shows the share of manufacturing in each location. It looks very much like Figure 1.5, which showed a run of the edge city model. Again, we see an almost flat surface begin to undulate, then gather itself up into two dorsal fins.

Given this picture, you will not be surprised to find that the same approach that allowed us to understand the origins of order in the edge city model works here as well. We can linearize the model around the unstable equilibrium in which manufacturing is evenly

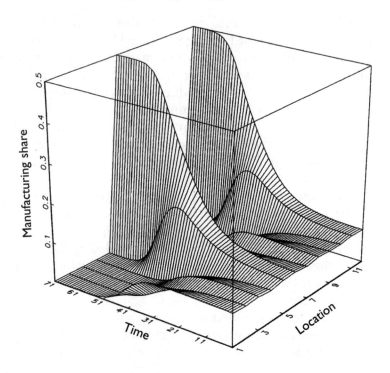

Figure 8.5 Evolution of Central Places. The dynamics of a central place
model show the same "'59 Cadillac" picture as the edge city
model.

spread across the landscape and represent the deviation of the spatial
distribution from flatness as a Fourier series. It then turns out that
there is a particular frequency of fluctuation that grows more rapidly than any other and tends first to dominate the distribution then
to gather itself up into spikes.

What may puzzle you, however – it puzzled me at first – is why
the model so nicely produces multiple agglomerations. In the edge
city model, the key to the result was the assumption that centripetal
forces fall off more rapidly with distance than centrifugal forces. We
made no such assumption here. So what about the model creates an
equivalent effect, and is it plausible?

I can offer a heuristic explanation that suggests that multiple
agglomerations are indeed a natural outcome in a model in which
centripetal forces arise from market linkages. Suppose that instead
of realistically representing transportation costs as a steadily increas-

ing function of distance, we imagine instead a world in which transport costs are zero up to some distance, then infinite thereafter. Say this distance is 50 miles. Then the forward and backward linkages, which are the centripetal forces in the model, will extend only 50 miles.

But what about the centrifugal force – the competitive pressure that firms exert on each other? Well, even if manufactures can be shipped only 50 miles, any firm that is less than 100 miles away will still be in a position to compete for some of your customers. So the range of centrifugal forces, of the negative spillovers from the presence of rival firms, will indeed extend farther than the range of the positive spillovers from larger markets. And that is all that we need to get multiple agglomerations.

At this point let us pause and ask what we have just developed a model of. The answer, I think, is that we have just seen how a central place system along the lines envisaged by Lösch can emerge via the invisible hand. The essence of central place theory is the trade-off between economies of scale, which provide an incentive to concentrate production in a limited number of sites, and transportation costs, which provide an incentive to multiply sites so as to stay close to consumers. Lösch argued that to make the best of such a tradeoff one should have a regular lattice of central places (with hexagonal market areas in a two-dimensional economy). He was unable, however, to show how such a lattice might emerge through a decentralized, unplanned process. Well, here we have an example of just such a process. I have demonstrated the emergence of a regular lattice only for a one-dimensional economy, but I have no doubt that a better mathematician could show that a system of hexagonal market areas will emerge in two dimensions.

Moreover, it seems to me that we can adapt central place theory the same way that Edwin Mills adapted the von Thünen model: we can recast it as a model of the *internal* structure of urban sprawls, with suburban families as farmers and office buildings as factories. I do not want to go through the details of such a reinterpretation, but I am sure you get the idea. The important point is that we see how the mysterious centripetal and centrifugal forces of the edge city model could arise from pecuniary externalities, alias market-size effects. People like Philip Anderson should be gratified: in this view, the agglomeration economies that drive the formation of edge cities are entirely an emergent property, with not even a hint of such a property at the level of individual firms.

So I would offer a double reinterpretation of this model that started out as being about farmers and manufacturers. First, it implies the formation of Lösch-type central places and may thus be taken as a formalization and vindication of central place theory. Second, as an empirical matter, it can be interpreted as a model of what happens inside metropolitan areas as well as (or indeed in preference to) a model of the formation of such areas themselves.

SIMON'S URBAN GROWTH MODEL

Finally, let me turn to the very different story of urban growth proposed by Herbert Simon to explain the stunning empirical success of a power law in describing city size distributions. I have some serious problems with this model; but it is a thing of beauty and also the best explanation anyone has yet proposed, so let me set out the model before criticizing it.

How Random Growth Produces a Power Law

Recall again Simon's story. It is one of lumps and clumps: as each new "lump" of population arrives, it either starts a new city (with probability π) or attaches itself to an existing "clump"; the probability that each clump attracts the lump is proportional to that clump's population. Thus we have a process of random growth of clumps, with the expected rate of growth independent of clump size.

To see why this story works, let me introduce some notation. Let S be city size, measured in "lumps"; let P be the population, also measured in lumps; let N_s be the number of cities (clumps) of size S or larger; and let n_s be the number of cities of precisely size S. If city sizes obey a power law, then N_s is proportional to $S^{-\alpha}$, and n_s is proportional to $S^{-\alpha-1}$. Or to put it in a different and useful way, we want a story that implies an *elasticity* of n with respect to S of $-\alpha-1$.

It is natural, though not crucial, to suppose that the population grows at an expected constant rate γ through Poisson addition of lumps.

To analyze the model, we make a crucial guess: that the urban system tends to approach a steady state, in which the number of cities of size S reaches a constant ratio to the population. This cannot, of course, ever be exactly true, because the largest city will always tend to grow larger over time. But let us disregard this for a moment and suppose that n_S/P does in fact tend to approach a constant value for all S.

How does n_S/P change? The denominator grows with the general growth in the population. The numerator can change for two reasons. First, a "lump" can attach itself to a city of size $S - 1$, turning it into a city of size S and adding 1 to n_S. Second, a lump can attach itself to a city of size S, turning it into a city of size $S + 1$ and reducing n_S by 1. But recall that with a probability π an arriving lump forms a new city. If it does attach itself to an existing city, then it has a probability $n_S S/P$ – the combined population of cities of that size, divided by the total urban population – of attaching itself to a city of size S. Writing the expected rates of change and fudging a bit over the discontinuous growth of P, we have

$$\frac{E[d(n_s/P)]}{dt} = \frac{\gamma}{P}\Big[(1-\pi)n_{s-1}(S-1) - (1-\pi)n_s S - n_s\Big] \qquad (8.4)$$

But if the system approaches a steady state, the right-hand side of (8.4) must approach 0. This implies that in the steady state we must have

$$\frac{n_s}{n_{s-1}} = \frac{(1-\pi)(S-1)}{(1-\pi)S + 1} \qquad (8.5)$$

This may be rewritten as

$$\frac{n_s - n_{s-1}}{n_{s-1}} = \frac{\pi - 2}{(1-\pi)S + 1} \qquad (8.6)$$

If we restrict our attention to large cities, we may approximate the discrete distribution of city sizes by a continuous one; in particular, it will be approximately true that

$$n_s - n_{s-1} = \frac{dn}{dS} \qquad (8.7)$$

Substituting this approximation back into (17), we have an expression for the elasticity of n with respect to S:

$$\frac{dn}{dS}\frac{S}{n} = \frac{\pi-2}{1-\pi+1/S} \qquad (8.8)$$

However, when S is large – that is, when we look at the upper tail of the city size distribution – this simplifies to

$$\frac{dn}{dS}\frac{S}{n} = \frac{\pi-2}{1-\pi} = -\alpha-1 \qquad (8.9)$$

And so in the end we have the implied exponent on the power law:

$$\alpha = \frac{1}{1-\pi} \qquad (8.10)$$

It is a remarkably elegant result. And it seems to suggest a very simple explanation not only for the existence of a power law, but for the exponent of 1: new units of economic activity almost always form in existing clusters.

I believe that this insight must be essentially right – although some problems with Simon's story still give me sleepless nights. I shall come back to those problems in a moment. First, however, let me point out some implications of Simon's model that may not be entirely obvious.

Implications of the Model

The model that Simon suggested to explain the city size distribution is in sharp contrast to the models I have suggested to explain the formation of central places. Central place theory depends crucially on economies of scale – on external economies in particular, although these external economies may be an emergent consequence of internal scale economies. But I have suggested that power laws typically arise from processes of random growth that are approximately scale independent. How can these views be reconciled? Does the evidence that city sizes follow a power law imply that there are no economies of scale in city formation?

Herbert Simon seems to have thought so. Indeed, Simon's disenchantment with standard economics seems to have been driven in part by his observation that firm sizes, like city sizes, follow a power law, and his conclusion that the U-shaped average cost curve of the textbooks was nonsense. Yet a closer consideration of his own model suggests that it is not consistent with strictly constant returns.

To see why, suppose that there were hardly any economies of scale. This would imply, in Simon's story, that the "lumps" of population would be very small. But suppose that the lumps were very small relative to the size of most cities. Then the law of large numbers would apply: all cities would grow at the same rate – we would have growth, but not random growth – and whatever size distribution we started with would be preserved. In particular, if cities started out of roughly equal size, there would be no way to get to the wide range of inequality implied by power laws with small exponents. To get convergence to a power law, then, the lumps must not be too small. Or, to be more precise, the smaller are the lumps, the greater the percentage population increase needed to produce a smooth power law. So the fact that the United States has actually achieved a quite smooth power law distribution since it stopped generating lots of new cities – say, since 1860 – is an indication that the lumps cannot be too small.[6]

On the other hand, the lumps cannot be too large, or it would be impossible to get a smooth distribution of city sizes in any case. The rank–size rule works spectacularly well for American cities over a range of roughly two orders of magnitude, from cities of around 200,000 up to metropolitan New York, with almost 20,000,000.

---

6. In the simulation shown in Figure 3.3, I allowed the urban population to grow by a factor of 100. Is this a reasonable number? In reality, America stopped forming major new cities east of the Rockies around 1860, when the so-called manufacturing belt locked into place, a development that I described in *Geography and Trade* and attributed to railroads, manufacturing scale economies, and the diminishing importance of agriculture. Since then the urban population has increased by a factor of only 20. The 1860 city size distribution was not, however, a set of equal size "seeds," like the initial condition I used in my simulation. Instead, it was a highly unequal clumping, with big cities at strategic locations like the mouth of the Hudson. And I am reasonably sure that convergence to a smooth power law should take place faster when one starts from a suitably uneven initial condition. I like to imagine that someday, when we are more sure about the origins of order in urban systems, economists will be in a position to ask whether any given theory is sufficient to explain the evolution of order in the time available, and even to use that evolution to make inferences – just as cosmologists have used the observed large-scale structure of the universe to infer the existence and nature of "dark matter."

This suggests that the size of a lump – in effect, the minimum efficient scale for a city – cannot be more than about 200,000 people. (For what it is worth, the biggest city in Figure 3.3, which shows a run of the Simon model with $\pi = 0.2$, has 141 lumps.)

There is an enormously tempting possibility here: to identify the lumps that make up metropolitan areas with Garreau's edge cities. If we could do this, we could nicely reconcile central place models that rely on increasing returns – and the anecdotal evidence that suggests substantial increasing returns at the level of urban subcenters – with the more or less constant returns suggested by the metropolitan size distribution. I have my doubts about whether we can actually make this fit – it looks to me as if edge cities are too big to do the job. After all, a power law really should work well only for the upper tail of the city size distribution; so the smallest cities for which it works should actually consist of several lumps. Yet Garreau suggests that an edge city normally services a population of at least 250,000, which is a bit bigger than the smallest cities that still obey the rank–size rule. The fit does not seem right – yet it is close enough to be tantalizing.

In any case I have a problem with this whole analysis, which I may as well share.

An Unresolved Problem

In the simulation run shown in Figure 3.3, I set the probability of forming new cities at 0.2 and got a power law distribution with a slope fairly close to –1. But the real city size distribution has a slope that is almost exactly 1, and Simon's model suggests that this is what we should expect with $\pi = 0$. Why didn't I run the model with $\pi = 0$?

Because it doesn't work, that's why. Try it if you are a masochist (I wasted quite a few days trying to find the error in my program before realizing that it was not a computer problem). No matter how long you run a simulation of Simon's model with all lumps attaching themselves to existing clumps, you will never get a log–linear relationship between city size and city rank.

The reason is hidden in the logic of the model. To justify the emergence of a power law, Simon assumes that the size distribution of firms approaches a steady state. As I have already pointed out, this never happens completely, because the biggest city keeps on getting bigger. But as long as at least some new cities are formed, this turns out not to be a problem. The reason is that, although the biggest cities keep getting bigger, very big cities eventually hold a trivial share of the total population, so that almost all population growth takes place within existing size classes.

Let us be a bit more formal (although this argument is far from rigorous, as it is slightly circular). Suppose that the size distribution really were a power law throughout, with the number of cities bigger than S proportional to $S^{-\alpha}$, and with $S_{\min}$ the smallest possible city. Then the share of population in cities larger than any given S would be $(S/S_{\min})^{1-\alpha}$. As long as $\alpha > 1$, really big cities will have a trivial share of the population and will therefore attract hardly any of the growth in population; so it is reasonable to think of cities below some large size as approximately reaching a steady state once the biggest cities have grown big enough.

But what if α equals 1, as it should if no new cities form (and as it does in fact in U.S. data)? Then we have an absurd result: if the power law really held, all population increase should occur in cities of a size greater than S, *no matter how large* S *is*. This is nonsensical – what it really means is that the biggest cities keep getting bigger, and their growth involves so much of the population increase that the distribution never gets to look anything like a power law.

You might say that this is only a limiting case; let us just imagine that almost all, rather than actually all, lumps attach themselves to existing clumps. But this is not really an answer, because when α is very close to 1, the model will not start to approach a steady-state distribution until the biggest cities are very big indeed, which would not happen unless the process of random growth has gone on a long time. To get the smooth result in Figure 3.3, I had to let the urban population increase by a factor of 100, which is just barely reasonable for the United States. This, however, was for a simulation in which the probability of new city formation was 0.2. If I had tried to set that probability any lower, to reproduce the incredible exactness with which α seems to equal 1, I would have needed a far larger population increase.

At the moment, I have no resolution for this problem. Nonetheless, I remain strongly attached to two propositions. First, the power law on city sizes is very real and tells us something very important about our economy. Second, some kind of random growth process is far and away the most likely explanation. And Simon's model is so wonderfully elegant an approach that I continue to regard it as the best game in town.

——————————9——————————

Concluding Thoughts

The world is full of self-organizing systems, systems that form structures not merely in response to inputs from outside but also, indeed primarily, in reponse to their own internal logic. Global weather is a self-organizing system; so, surely, is the global economy.

Are there universal laws governing self-organizing systems? I do not know, and I am surely not the person to find them. There are, however, some characteristic ways in which systems are known to organize themselves, and scientists have become increasingly aware that the existence of these characteristic styles of self-organization leads to parallels between phenomena that might seem completely distinct. The physicist John Hopfield argued, in a now famous paper, that the spin-glass models of condensed matter physics could shed considerable light on how memory and learning take place in the brain; it should not then surprise us that Schelling's model of segregation looks a lot like a spin-glass, too. Per Bak has argued that similar random growth processes can explain the size distributions of everything from earthquakes and forest fires to species extinctions; it is not unreasonable to suppose that something along the same lines may apply to city sizes as well.

In this book I have suggested two principles of self-organization that seem to me to be particularly useful in explaining economic behavior. The first principle is that of *order from instability*: when a system is so constituted that a flat or disordered structure is unstable, order spontaneously emerges. The classic physical example of this principle is convection; the global atmospheric circulation, which is primarily a giant convection machine, spontaneously organizes itself into weather patterns, essentially because a smooth circulation of air from poles to equator and back again is unstable. The business cycle theorists of the 1940s and 1950s thought that economic fluctuations arose out of a similar logic. I am not sure whether they were right, but I am convinced that this is in fact the right way to think about how metropolitan areas evolve their far from uniform structure.

The second principle is that of *order from random growth*. Objects of many kinds, from earthquakes to asteroids, obey a power law size distribution. The best explanation is that these objects are formed by a growth process in which the expected rate of growth is approximately independent of scale, but the actual rate of growth is random. The size distribution of cities exhibits sustained empirical regularities that are every bit as striking and consistent as those of the physical processes.

What good is the idea of a self-organizing economy? Well, anything that makes us reconsider the way we think about economics is bound to have policy relevance, but at this point I have no recommendations to offer. I can see vaguely how my spatial models could help in urban planning. To be honest, I can see somewhat more clearly how they could be helpful in real estate speculation! If something like the Scheinkman–Woodford model turns out to be really useful in looking at economic fluctuations, it could similarly be useful both for policy and for forecasting.

Luckily for me, a set of speculative discourses like this need not provide anything of immediate use. I hope that you have found the idea of the self-organizing economy as interesting and exciting as I do. More to the point, I hope that I have convinced you that the crazy things I have been saying might even be true.

————10————

Appendix:
The Evolution of Central Places

Discourses are no place for dense algebra. Yet the idea of self-organization is essentially a mathematical one and needs to be backed up by some reasonably serious algebraic grinding. So here are the models underlying what I have said about edge cities and central place theory.

URBAN MORPHOGENESIS: THE EDGE CITY MODEL

The objective of this model is to illustrate in as simple a fashion as possible how interdependent location decisions by individual businesses can lead to a self-organization of space. The basic idea is that there is a tension between "centripetal" forces that pull businesses together and "centrifugal" forces that drive them apart. What we want to see is how, and under what conditions, this tension can lead to the kind of polycentric pattern that characterizes modern metropolitan areas.

We simplify the problem by assuming a one-dimensional metropolitan area. It will be useful to move back and forth between explicitly representing a metropolitan area as a circle, with a circumference normalized to 2π, and treating the metropolitan area as if it were of infinite extent – which may simply be seen as an approximation to self-organization on a very large circle. The analytics of the infinite metropolis (I guess that is Los Angeles) are somewhat simpler and more intuitive, but the numerical examples must focus on the slightly harder finite city case.

We begin by specifying the interdepedence between firm locations. Let x be some location on the line, and let $\lambda(x)$ be the *density* of firms at that location. We assume that the desirability of any given location – which, following the tradition of Harris (1954) and Lowry (1965) we can call the location's *market potential* – depends both positively and negatively on the density of firms at other locations. Both forces decline with distance, but the positive forces decline faster. In particular, we assume that the market potential equation takes the form

$$P(x) = \int_z [A \exp(-r_1 D_{xz}) - B \exp(-r_2 D_{xz})]\lambda(z)dz \qquad \text{(A.1)}$$

where D_{xz} is the distance between x and z, and where $r_1 > r_2$. Here, A represents the strength of the centripetal, agglomerative forces; B, the strength of the dispersing, centrifugal forces; and r_1 and r_2 represent the rates at which these forces dissipate with distance. By making $r_1 > r_2$, we introduce the essential difference in range that makes multiple agglomerations possible.

It seems reasonable to suppose that agglomeration will take place only if A is sufficiently large relative to B, but that multiple subcenters are possible only if B is sufficiently large relative to A. We will see shortly that this is correct; indeed, the interesting range is where

$$\frac{r_1}{r_2} > \frac{A}{B} > \frac{r_2}{r_1} \qquad \text{(A.2)}$$

To complete this minimalist model, we need only add some dynamics. Let us assume that businesses gradually migrate toward locations with above-average market potential P and away from those with below-average potential. The average market potential may be defined as

$$\overline{P} = \int_x P(x)\lambda(x)dx \qquad \text{(A.3)}$$

A simple dynamic rule that obeys the adding-up constraint that the total number of businesses remain constant is

$$\frac{d\lambda(x)}{dt} = \gamma[P(x) - \overline{P}]\lambda(x) \qquad \text{(A.4)}$$

Notice that the $\lambda(x)$ in equation (A.4) serves two purposes. It has to be there to ensure that the sum of changes add to zero. It also ensures that the density at any point will never fall below zero, so that we need not introduce the requirement that there be a non-negative number of businesses at each location as an explicit constraint.

Unfortunately, the necessary presence of this term also implies that the model is nonlinear, as indeed the nonnegativity constraint also does. So in spite of the simplicity of the model's description, we seem to be dealing with a nonlinear dynamic model with a continuum of variables. How is such an apparent monster to be analyzed?

One answer is to turn to the computer. It is a simple matter to set up a circular model with a fairly large number of discrete locations, start with a random distribution of business across these locations, and simply see what happens for a number of values of the parameters. Figures 1.5 and 1.6 show sample results for a 24-location model. (It is a good idea to let the number of locations have a large number of divisors.) One might then look for "empirical" regularities in the results of these "experiments."

However, the experimental mathematics is so successful at yielding regularities that we are driven back to to pencil and paper! For any given set of parameters, the model produces not only multiple centers but a consistent number of centers, more or less evenly spaced around the circle. This surprising predictability of behavior becomes even more striking when we choose a rule for setting initial business shares that yields a fairly even distribution around the circle. For Figures 1.5 and 1.6, the share i of business at any location i was set equal to

$$\lambda_i = \frac{k + u_i}{\sum_j (k + u_j)} \tag{A.5}$$

where u_i was a drawing from the uniform distribution between 0 and 1, and k was a "smoothing" parameter; the larger is k, the closer the initial distribution will be to flat. And, once k is set fairly high (it was 5 for those figures), one consistently gets the same number of equally spaced, equal size centers for any given parameters.

How is this result to be explained? By focusing not on the whole path to long-run equilibrium but on the process of divergence away from a flat, uniform spatial structure.

Linear Dynamics: The Large City Case

By the *large city case* I mean a city sufficiently large that we can think of it as if it were a line of infinite length. This means ignoring two annoying details that arise when you deal with finite size circles: that the distance between two points can never exceed half the circumference, and that the shortest route to another point may involve going either clockwise or counterclockwise. In the large city case we integrate over z as if it ranged from minus to plus infinity and measure distances by

$$D_{xz} = |x - z| \qquad\qquad (A.6)$$

Suppose, now, that we consider the behavior of the model defined by equations (A.1), (A.3), and (A.4) not in general, but only in the neighborhood of a completely uniform spatial distribution of business — $\lambda(x) = \bar\lambda$ for all x. (It is helpful to choose units so that at this uniform density $\bar\lambda(x) = 1$.) In this case the model can be given a linear approximation and can indeed be collapsed into a single (partial differential) equation:

$$\frac{d\lambda(x)}{dt} = \gamma \int_{-\infty}^{\infty}[Ae^{-r_1|x-z|} - Be^{-r_2|x-z|}](\lambda(z) - 1)dz \qquad (A.7)$$

Let $\lambda'(x) = \lambda(x) - 1$. Then it is a general principle that this deviation of the business density from the average can be represented as the sum of a number of periodic fluctuations at different frequencies:

$$\lambda'(x) = \textstyle\sum_i \lambda'_i(x) \qquad\qquad (A.8)$$

where

$$\lambda'_i(x) = a_i \sin(\phi_i x) + b_i \cos(\phi_i x) \qquad\qquad (A.9)$$

Now imagine for the moment that the distribution of business was completely described by a fluctuation at a particular frequency ϕ; and suppose without loss of generality that this fluctuation had a peak at $x = 0$, so that we could simply write

$$\lambda'(x) = h(t)\cos(\phi x) \qquad\qquad (A.10)$$

Substitute this back into (A.7), and we get a rather simple expression for the change in the density:

$$\frac{d\lambda'(x)}{dt} = \gamma\left[A\,\frac{r^1}{r_1^2+\phi^2} - B\,\frac{r^2}{r_2^2+\phi^2}\right]\lambda(x) \qquad (A.11)$$

That is, if the density could be described by a fluctuation at a frequency ϕ, then that fluctuation would grow at a rate $g(\phi)$.

This may seem a moot point, but it is not, because if the actual deviation of the density from flatness can be described by a sum of sinusoidal fluctuations, then the growth of that deviation can be thought of as the sum of the fluctuations multiplied by their characteristic growth rates:

$$\frac{d\lambda'(x)}{dt} = \Sigma_i g(\phi_i)\lambda'_i(x) \qquad (A.12)$$

Therefore, we can think of the evolution of this spatial economy as consisting of the parallel growth of a set of hypothetical spatial economies, each of which is characterized by a regular distribution of business with a different wavelength.

Now suppose that the initial distribution of business density was very flat. As the fluctuations grow over time, the deviation of the density from flatness will not only increase but tend increasingly to be dominated by whatever frequency fluctuation grows fastest. So the crucial question is the shape of $g(\phi)$.

From inspection of equation (A.11), three things about $g(\phi)$ should be apparent. First, at very high frequencies, $\phi\to\infty$, $g(\phi)$ approaches zero. That is, very high frequency fluctuations go nowhere in particular. Second, at very low frequencies, $\phi\to0$, $g(\phi)$ will become negative as long as $A/r_1 < B/r_2$. Finally, at sufficiently high ϕ, $g(\phi)$ will be positive as long as $Ar_1 > Br_2$. Thus as long as the criterion (A.2) is satisfied, $g(\phi)$ should be negative at low frequencies, approach zero at high frequencies, and reach a positive maximum somewhere in between.

The Fourier series representation of a distribution along a very long line will involve many different frequencies; in effect, the spectrum will be covered quite closely. So in the infinite city case, we can simply think of the deviation of the economy from flatness as being dominated over time by whatever frequency maximizes the function $g(\phi)$. The wavelength that corresponds to that frequency is the normal distance between edge cities!

The Small City Case

For simulation analysis, it is necessary to work with a "small city": a circular array of locations with a distance small enough that only a few concentrations develop. It is convenient, in fact, to normalize units so that the circle is of circumference 2π.

When one works with the linearized version of such a small city model, two complications appear. First, only fluctuations that go an integer number of times around the circle are possible. For a circle of circumference 2π, this means that the allowed frequencies are 1, 2, 3, . . . Second, the maximum distance of any location from any other is π. This makes the expression for the growth rate of a regular fluctuation somewhat more complicated. It now takes the form

$$g(\phi) = \gamma\left[\frac{A}{r_1} H_1 - \frac{B}{r_2} H_2\right] \qquad (A.13)$$

where

$$H_i = \left[1 - (-1)^\phi e^{-r_i\pi}\right] \frac{r_i^2}{r_i^2 + \phi_2} \qquad (A.14)$$

(You can see from this where some of the odd sign reversals in Table 8.1 came from.)

The simulations for Figures 1.5 and 1.6 were both done with $A = 0.2$, $B = 1.0$. For Figure 1.5 and the first column of Table 8.1, the rates of decline were set at $r_1 = 1.4$, $r_2 = 0.2$; for Figure 1.6 and the second column of Table 8.1 they were set at $r_1 = 2.8$, $r_2 = 0.4$.

A CENTRAL PLACE MODEL

In several recent papers (Krugman 1991, 1993a, 1993b) I have explored one particular approach to spatial modeling that, although admittedly capturing only some of the reasons why spatial structure emerges in real economies, has the virtue of being particularly easy to work with. In this approach, an economy with two or more locations is assumed to consist of two sectors: a constant returns, geographically immobile sector ("agriculture"), and an increasing returns, monopolistically competitive, geographically mobile sector ("manufacturing"). When one adds transportation costs in the manufacturing sector and some simple dynamics, models of this type exhibit spontaneous spatial self-organization: even if all locations are identical in resources and technology, manufacturing firms have an incentive to concentrate production close to the markets and supplies that other manufacturing firms provide, thus producing a "centripetal" tendency toward agglomeration. Working against this centripetal tendency, however, is the "centrifugal" pull of the immobile agricultural sector.

In a two-location model, the tension between centripetal and centrifugal forces can be treated analytically; one can derive a criterion, depending in an economically meaningful way on the parameters, that determines whether or not manufacturing concentrates in one location. Beyond this case, however, it becomes very difficult to derive analytical results. Simulations show that there may be equilibria with multiple manufacturing concentrations; they also indicate that, as the number of locations grows, there typically start to be a very large number of equilibria.

And yet, underlying this complexity seems to be some order. When one starts from a random distribution of manufacturing on a linear landscape, for example, one typically finds that a roughly regular spacing of manufacturing concentrations emerges. Furthermore, the distance between these concentrations is relatively insensitive to the starting position and appears to depend in a sensible way on the model's parameters.

Given the results of the edge city model, these regularities are no longer surprising. One can immediately guess that they emerge from the process of divergence away from a flat equilibrium, and that in a linearized approximation to the model in the vicinity of a flat distribution, there is a particular frequency of fluctuation that tends to grow fastest. This turns out to be correct, but the story is a bit more complicated, because the centripetal and centrifugal forces are now derived rather than assumed.

I begin with a review of the general approach to spatial dynamics used here, then turn to a specific model of a linear economy and show how the evolution of this model near an even distribution of manufacturing can be viewed in terms of the growth rates of fluctuations of different frequencies. Finally, I show why the model economy has a preferred wavelength, and how this wavelength depends on the parameters.

A Basic Spatial Model

Consider an economy in which there are a number of locations, indexed by $j = 1, \ldots, J$. Let D_{jk} be the distance between any pair of locations j and k.

In this economy there are two factors of production: immobile "farmers" and mobile "workers." It will be convenient to choose units so that there are a total of $1 - m$ farmers and m workers. Also, in this appendix I will restrict attention to economies in which spatial structure is completely endogenous, so the farmers will be assumed to be equally divided among the locations.

Everyone in this economy shares the same tastes, which may be represented by a two-level structure. At the upper level, there are Cobb–Douglas preferences between agricultural goods and a manufacturing aggregate:

$$U = C_M^{\mu} C_A^{1-\mu} \tag{A.15}$$

At the lower level, manufacturing is a CES composite of a large number of symmetric differentiated products:

$$C_M = [\Sigma_i c_i^{\rho}]^{1/\rho} \tag{A.16}$$

where $\sigma = 1/(1 - \rho)$ is the elasticity of substitution.

Each factor is specific to the production of one sector. Farmers produce agricultural output with constant returns to scale. Workers produce manufactured goods. There are economies of scale in this production, specific to both the firm and the particular variety produced; these are represented as a linear cost function,

$$L_{Mi} = \alpha + \beta Q_{Mi} \tag{A.17}$$

We also introduce transport costs. For the sake of tractability, there are assumed to be zero transport costs for agricultural goods. Transport costs on manufactured goods are of Samuelson's "iceberg" form. If one unit of a manufactured good is shipped from location j to location k, only $\exp(-\tau D_{jk})$ units arrive, with λ the transportation cost per unit distance.

It is a familiar proposition that, if we take the spatial distribution of workers as given, a model of the form just described yields a monopolistically competitive equilibrium in which all profits are competed away. This equilibrium includes an equilibrium level of the real wage at each location; differences in these real wage rates are what drive the economy's dynamics.

Workers are assumed to move gradually toward locations that offer them above average real wages. Let λ_j be the fraction of workers currently in location j. Then the average real wage rate can be defined as a weighted average of real wage rates at each location,

$$\bar{\omega} = \Sigma_j \lambda_j \omega_j \tag{A.18}$$

and the assumed dynamics take the form[1]

$$\frac{d\lambda_j}{dt} = \gamma(\omega_i - \bar{\omega})\lambda_j \tag{A.19}$$

The dynamic behavior of this model can be thought of as a sequence of general equilibrium problems. For any given distribution of manufacturing across locations, the economy reaches an equilibrium that determines the real wage at each location. This vector of real wages then determines, via (A.18) and (A.19), the distribution of workers a short time later, and the calculation can be repeated until the model economy converges on some long-run equilibrium geographical pattern.

In Krugman (1992) I show that the equilibrium of this model at any point in time can usefully be described as the simultaneous solution of four sets of equations. First, the income of any location is the sum of the earnings of its immobile farmers and the workers who are currently located there:

$$Y_j = \frac{1-\mu}{J} + \mu\lambda_j w_j \tag{A.20}$$

---

1. In all of my models to date, I have ignored two important aspects of real-world spatial economics – forward-looking behavior by agents who try to anticipate future spatial patterns and large agents, such as shopping mall developers, who try to influence these patterns. The excuse for these omissions is, of course, tractability.

where w_j is the wage rate measured in terms of the agricultural good.

Second, the true price index of manufactures at any given location depends on the distribution of manufacturing, transportation costs, and wage rates:

$$T_j = \left[\sum_k \lambda_k w_k^{1-\sigma} e^{-\tau(\sigma-1)D_{jk}}\right]^{1/(1-\sigma)} \tag{A.21}$$

Third, the equilibrium wage rate at any location depends on incomes, true price indices, and transportation costs to all other locations:

$$w_j = \left[\sum_k Y_k T_k^{\sigma-1} e^{-\tau(\sigma-1)D_{jk}}\right]^{1/\sigma} \tag{A.22}$$

Finally, the real wage rate at location j depends on the nominal wage rate in terms of agricultural goods and the local true price index of manufactured goods:

$$\omega_j = w_j T_j^{-\mu} \tag{A.23}$$

These equations are fairly simple and very easy to solve numerically – one simply starts with guesses at the wage and true price vectors and iterates until convergence. Analytical results, however, in anything larger than a two-region model are another matter. Hence, explorations of multilocation settings so far have relied on numerical examples. Using the Fourier series approach, however, we can show the underlying logic of self-organization.

Dynamics Near a Flat Spatial Structure

For the formal analysis, we will consider a version of the basic model in which farmers are distributed evenly along a line of infinite extent. Workers will also, at any point in time, be distributed along that line; we let $\lambda(x)$ be the density of workers at position x, normalized so that with a flat distribution $\lambda = 1$ everywhere.

For this economy, equations (A.6)–(A.9) may be rewritten in the following form (the constant terms are added so that when the distribution is flat, $Y(x) = w(x) = T(x) = 1$ for all x is a solution):

$$Y(x) = 1 - \mu + \mu\lambda(x)w(x) \tag{A.24}$$

$$T(x) = \left[\frac{\tau(\sigma-1)}{2} \int_{-\infty}^{\infty} \lambda(z) w(z)^{1-\sigma} e^{\tau(1-\sigma)|x-z|} dz\right]^{1/(1-\sigma)} \tag{A.25}$$

$$w(x) = \left[\frac{\tau(\sigma-1)}{2} \int_{-\infty}^{\infty} Y(z) T(z)^{\sigma-1} e^{-\tau(\sigma-1)|x-z|} dz\right]^{1/\sigma} \tag{A.26}$$

$$\omega(x) = w(x) T(x)^{-\mu} \tag{A.27}$$

This set of nonlinear equations is fairly nasty looking. Suppose, however, we restrict our attention to situations in which $\lambda(x)$ is close to 1; that is, where the distribution of manufacturing is fairly flat. Then we can take linear approximations to the equations. Let a prime on a variable represent deviation from 1; then the approximate linearized model takes the form

$$Y'(x) = \mu\lambda'(x) + \mu w'(x) \tag{A.28}$$

$$T'(x) = \frac{1}{1-\sigma} \frac{\tau(\sigma-1)}{2} \left[\int_{-\infty}^{\infty} \lambda'(z) e^{-\tau(\sigma-1)|x-z|} dz + \int_{-\infty}^{\infty} w'(z) e^{-\tau(\sigma-1)|x-z|} dz\right] \tag{A.29}$$

$$w'(x) = \frac{1}{\sigma} \frac{\tau(\sigma-1)}{2} \left[\int_{-\infty}^{\infty} Y'(z) e^{-\tau(\sigma-1)|x-z|} dz + \int_{-\infty}^{\infty} T'(z) e^{-\tau(\sigma-1)|x-z|} dz\right] \tag{A.30}$$

$$\omega'(x) = w'(x) - \mu T'(x) \tag{A.31}$$

These equations do not, at first sight, appear any more tractable than the nonlinear version. But let us assume for a moment that the distribution of manufacturing follows a simple periodic distribution, say,

$$\lambda'(x) = \delta\cos(\phi x) \tag{A.32}$$

Now let us simply guess that, if the divergence of $\lambda(x)$ from 1 follows this simple periodic form, the divergences of all of the other variables from 1 will be constant multiples of $\lambda(x)$. (This conclusion is actually obvious from the spatial symmetry and the linearity.) That is, we guess that there is a solution of the form

$$Y'(x) = a_Y \lambda'(x) \tag{A.33}$$

$$T'(x) = a_T \lambda'(x) \tag{A.34}$$

$$w'(x) = a_w \lambda'(x) \tag{A.35}$$

$$\omega'(x) = a_\omega \lambda'(x) \tag{A.36}$$

If this is a valid solution, then we have managed to reduce a general equilibrium problem that is, strictly speaking, the solution of an infinite number of nonlinear equations to the solution of four linear equations.

Let us, then, substitute (A.32) into (A.33)–(A.36). When we do so, we will see repeatedly a term of the form

$$K(z) = \frac{\tau(\sigma-1)}{2} \int_{-\infty}^{\infty} \cos(\phi z) e^{-\tau(\sigma-1)|x-z|} dz \tag{A.37}$$

With a little grinding, it is possible to show that

$$K(z) = H(\phi, \tau, \sigma)\cos(\phi z) \tag{A.38}$$

where

$$H(\phi, \tau, \sigma) = \frac{(\sigma-1)^2}{(\sigma-1)^2 + (\phi/\tau)^2} \tag{A.39}$$

H represents a sort of discount factor: the ratio of the impact of a fluctuation to what would happen if there were a uniform increase in the same variable that raised the level at x by the same amount. Therefore in the equation for the true price index we know that an equal increase in all wage rates would raise the price index at x by an amount equal to the increase in the wage rate at x; a fluctuation will raise the price index by *H* times the increase at x, with the ratio

H depending on the frequency of the fluctuation. It is immediately obvious that for very high frequencies, H approaches 0, whereas for low frequencies, it approaches 1.

We can now write our equations as

$$a_Y = \mu a_w + \mu \qquad (A.40)$$

$$a_\tau = \frac{-1}{\sigma - 1} H + H a_w \qquad (A.41)$$

$$a_\omega = \frac{1}{\sigma} H a_Y + \frac{\sigma - 1}{\sigma} H a_T \qquad (A.42)$$

$$a_\omega = a_w - \mu a_T \qquad (A.43)$$

These equations can be solved to yield the crucial result that

$$a_\omega = \frac{\mu}{\sigma - 1} H + (1 - \mu H) \frac{\mu H - H^2}{\sigma - (\sigma - 1 + \mu) H} \qquad (A.44)$$

Why is this the crucial result? Because the linearized version of the dynamic equation (A.19) is

$$\frac{d\lambda'(x)}{dt} = \omega'(x)\gamma = a_\omega \gamma \lambda'(x) = g_\phi \lambda'(x) \qquad (A.45)$$

where g is the rate of growth of a fluctuation at that frequency.

Now we note that a perturbation of the spatial distribution of manufacturing around $\lambda = 1$ can be represented as the sum of a number of sine waves of different wavelengths:

$$\lambda'(x) = \lambda'_1(x) + \lambda'_2(x) + \ldots \qquad (A.46)$$

And the growth of the perturbation may be written

$$\frac{d\lambda'(x)}{dt} = g_1 \lambda'_1(x) + g_2 \lambda'_2(x) + \ldots \qquad (A.47)$$

so that we can think of each periodic fluctuation as growing at its own characteristic rate. The fluctuation that will grow fastest is the one with the largest (positive) response of the real wage rate to manufacturing concentration and, given sufficient time, that fluctuation will dominate the spatial pattern.

So all we have to do to determine the preferred wavelength is find the maximum of (A.44). It is straightforward to determine three results. First,

$$a_\omega = 0 \text{ when } H = 0; \text{ that is, when } \phi \to \infty \qquad (A.48)$$

That is, fluctuations at very high frequencies – very short wavelengths will not tend to grow. Second,

$$a_\omega < 0 \text{ when } H = 1, \text{ provided that } \mu < \frac{\sigma - 1}{\sigma} \qquad (A.49)$$

The condition here is a familiar one, appearing also in Krugman (1991). It says, in effect, that economies of scale are not so large that all workers would prefer to be concentrated in the same place no matter how high transportation costs are. Given this condition, we find that very low frequency fluctuations, those with very long wavelengths, tend to die out.

Finally, at $H = 0$ we find

$$\frac{da_\omega}{dH} = \frac{\mu}{\sigma - 1} + \frac{\mu}{\sigma} > 0 \qquad (A.50)$$

Taken together, these observations imply that the relationship between the growth rate of a fluctuation and H is like an inverted J. Growth is slow at very short wavelengths, negative at high wavelengths, and most rapid at some intermediate wavelength.

The preferred wavelength, the wavelength of most rapid divergence, is a function of the three parameters τ, σ, and μ. The transport cost τ enters the solution in only one place, in the definition of H in (A.39). It is therefore obvious that the preferred frequency is strictly proportional to τ and hence that the preferred wavelength is inversely proportional. This is obvious with hindsight, because the wavelength and the transportation cost can both be changed in the same proportion by redefining the unit of distance, with no real change in the model.

It is more painful to derive the impact of changes in the elasticity of substitution and the share of manufacturing. It is, however, straightforward to calculate the preferred frequency numerically for given μ and σ. This is shown in Table A.1; we see that higher elasticities of substitution, which imply lower equilibrium economies of scale, tend to reduce the preferred wavelength, whereas a higher manufacturing share tends to increase the preferred wavelength.

Table A.1

σ	μ	.2	.3	.4
4		5.82	4.48	3.60
5		7.76	6.11	5.00
6		10.00	7.64	6.25

Preferred values of ϕ/τ in central place model

We see, then, that a model in which there are no assumed agglomeration economies or diseconomies, in which centripetal and centrifugal forces are entirely emergent properties, exhibits behavior that is very similar to that in a model in which these forces are simply assumed in a reduced form. And in both models, we find that the economy undergoes a systematic process of self-organization that can produce highly regular structures.

References

Arthur, W. B. (1994). *Increasing Returns and Path Dependence in the Economy*. Ann Arbor: University of Michigan Press.

Bak, P. (1991). "Self-Organizing Criticality." *Scientific American* (January).

Christaller, W. (1933). *Central Places in Southern Germany*. Jena: Fischer. English translation by C. W. Baskin. London: Prentice-Hall, 1966.

Cronon, W. (1991). *Nature's Metropolis*. New York: W. W. Norton.

Dicken, P., and Lloyd, P. (1990). *Location in Space*. New York: Harper and Row.

Dixit, A., and Stiglitz, J. (1977). "Monopolistic Competition and Optimum Product Diversity." *American Economic Review* (June).

Fujita, M. (1989). *Urban Economic Theory*. Cambridge: Cambridge University Press.

Fujita, M., and Ogawa, H. (1982). "Multiple Equilibria and Structural Transition on Non-Monocentric Urban Configurations." *Regional Science and Urban Economics* 12 (May): 161–196.

Garreau, J. (1992). *Edge City*. New York: Anchor Books.

Goodwin, R. (1951). "The Nonlinear Accelerator and the Persistence of Business Cycles." *Econometrica* 19: 1–17.

Harris, C. (1954). "The Market as a Factor in the Localization of Industry in the United States." *Annals of the Association of American Geographers* 64: 315–348.

Hicks, J. R. (1950). *A Contribution to the Theory of the Trade Cycle*. Oxford: Oxford University Press.

Hopfield, J. J. (1982). "Neural Networks and Physical Systems with Emergent Collective Computational Abilities." *Proceedings of the National Academy of Sciences* 79: 2554–2558.

Ijiri, Y., and Simon, H. (1977). *Skew Distributions and the Sizes of Business Firms*. Amsterdam: North-Holland.

117

Kauffman, S. (1993). *The Origins of Order*. New York: Oxford University Press.

Krugman, P. (1991). "Increasing Returns and Economic Geography." *Journal of Political Economy* (June).

Krugman, P. (1993a). "On the Number and Location of Cities." *European Economic Review* (May).

Krugman, P. (1993b). "First Nature, Second Nature, and Metropolitan Location." *Journal of Regional Science*.

Lewin, R. (1992). *Complexity: Life at the Edge of Chaos*. New York: Macmillan.

Lösch, A. (1940). *The Economics of Location*. Jena: Fischer. English translation, New Haven, Conn.: Yale University Press, 1954.

Lowry, I. S. (1966). *Migration and Metropolitan Growth: Two Analytical Models*. San Francisco: Chandler.

Mankiw, N. G. (1994). *Macroeconomics*, 2nd ed. New York: Worth.

Murphy, R., Shleifer, A., and Vishny, R. (1989). "Industrialization and the Big Push." *Journal of Political Economy*.

Nicolis, G., and Prigogine, I. (1989). *Exploring Complexity*. New York: W. H. Freeman.

Prigogine, I., and Stengers, I. (1984). *Order out of Chaos*. New York: Bantam Books.

Scheinkman, J. A., and Woodford, M. (1994). "Self-Organized Criticality and Economic Fluctuations." *American Economic Review* 84 (May): 417–421.

Schelling, T. (1978). *Micromotives and Macrobehavior*. New York: W. W. Norton.

Simon, H. (1955). "On a Class of Skew Distribution Functions." *Biometrika*.

Tobin, J. (1955). "A Dynamic Aggregate Model." *Journal of Political Economy* 63: 103–115.

Turing, A. (1952). "The Chemical Basis of Morphogenesis." *Philosophical Transactions of the Royal Society of London* 237: 37.

Waldrop, W. M. (1993). *Complexity*. New York: Basic Books.

Index

agglomerations, 33, 102, 107, 115; generated by central place model, 88-9; multiple, 75n1, 77, 88, 90-1; shadow, 87; temporal equivalent of, 68
Anderson, Philip, 3, 11, 91
Arthur, Brian, 87

Bak, Per, 44, 49, 69, 70, 99
Baumol, William, 9
Bjerknes, Vilhelm, 48
Blinder, Alan, 9
business cycle, 5, 6, 53, 61-3; global, 71-3; nonlinear theory of, 6, 7, 63-8, 71, 72-3, 100; order from random growth and, 69-71; power law for size distribution of, 70, 71

central place model, 76, 88-92, 106-15
central place theory, 13-15, 42, 76
central places: evolution of, 101-15; hierarchy of, 41-3
centripetal and centrifugal forces: business cycle, 68; central place model, 33, 88, 89, 90, 91, 107, 115; edge city model, 77, 81, 83, 101

change, process of: in economy, 56, 59; technology choice, models of, 57-9. *see also* dynamic systems
chaos, 48
Chicago, 13, 59
Christaller, W., 14-15
Citibank, 2
cities. *see also* edge cities; metropolitan areas; economics and, 9; explosive growth of, 59; Mills monocentric model of, 12-13; one-dimensional models of, 22, 77, 101; rank-size distributions. *See* rank-size rule, city
Cobb-Douglas preferences, 108
collective behavior, 3
competition for land, 10, 11-12, 13
complex landscapes, 31-8; of economy, 55-6; punctuated equilibrium and, 59; of technological change, 57
Complexity: Life at the Edge of Chaos, 6
Complexity: The New Science at the Edge of Order and Chaos, 66
complexity, theory of, 1-7. *see also* complex landscapes; economics,